PERGAMON INTERNATIONAL LIBRARY
of Science, Technology, Engineering and Social Studies

The 1000-volume original paperback library in aid of education,
industrial training and the enjoyment of leisure
Publisher: Robert Maxwell, M.C.

QUATERNARY GEOLOGY
A Stratigraphic Framework for Multidisciplinary Work

THE PERGAMON TEXTBOOK
INSPECTION COPY SERVICE

An inspection copy of any book published in the Pergamon International Library will gladly be sent to academic staff without obligation for their consideration for course adoption or recommendation. Copies may be retained for a period of 60 days from receipt and returned if not suitable. When a particular title is adopted or recommended for adoption for class use and the recommendation results in a sale of 12 or more copies, the inspection copy may be retained with our compliments. The Publishers will be pleased to receive suggestions for revised editions and new titles to be published in this important International Library.

Other Titles of Interest

ALLUM: Photogeology and Regional Mapping

ANDERSON: The Structure of Western Europe

ANDERSON & OWEN: The Structure of the British Isles

CONDIE: Plate Tectonics and Crustal Evolution

GRAY & LOWE: Studies in the Scottish Lateglacial Environment

OWEN: The Geological Evolution of the British Isles

ROBERTS: Geotechnology: An Introductory Text for Students and Engineers

QUATERNARY GEOLOGY

A Stratigraphic Framework for Multidisciplinary Work

by

D Q BOWEN

*University of Wales Reader in
the Department of Geography at the
University College of Wales, Aberystwyth*

PERGAMON PRESS

OXFORD · NEW YORK · TORONTO · SYDNEY · PARIS · FRANKFURT

U.K.	Pergamon Press Ltd., Headington Hill Hall, Oxford OX3 0BW, England
U.S.A.	Pergamon Press Inc., Maxwell House, Fairview Park, Elmsford, New York 10523, U.S.A.
CANADA	Pergamon of Canada Ltd., 75 The East Mall, Toronto, Ontario, Canada
AUSTRALIA	Pergamon Press (Aust.) Pty. Ltd., 19a Boundary Street, Rushcutters Bay, N.S.W. 2011, Australia
FRANCE	Pergamon Press SARL, 24 rue des Ecoles, 75240 Paris, Cedex 05, France
FEDERAL REPUBLIC OF GERMANY	Pergamon Press GmbH, 6242 Kronberg-Taunus, Pferdstrasse 1, Federal Republic of Germany

First edition 1978

British Library Cataloguing in Publication Data

Bowen, David Q
Quaternary geology.
1. Geology, Stratigraphic - Quaternary
I. Title
551.7'9   QE696   78-40288
ISBN 0-08-020601-8 Hardcover
ISBN 0-08-020409-0 Flexicover

In order to make this volume available as economically and as rapidly as possible the author's typescript has been reproduced in its original form. This method unfortunately has its typographical limitations but it is hoped that they in no way distract the reader.

*Printed in Great Britain by William Clowes & Sons Limited
London, Beccles and Colchester*

TO
MY MOTHER AND FATHER
AND ELIZABETH

CONTENTS

PREFACE

The discoveries of recent years have revolutionized the study of the Quaternary Period. To the student, and others on first acquaintance, these merely
appear to compound an already bewildering situation. The volume of often
seemingly contradictory literature continues to grow and its terminology to
multiply. Meanwhile terms long redundant for the specialist continue to
obtain currency and add to the confusion. Together with a certain degree of
particularism in the subject, wherein specialists tend to regard their fields
as the most definitive, it is not surprising that every kind of possible
misconception obtains. This is unfortunate because Gignoux's dictum that
'nothing excites the imagination more than the study of the Quaternary' is as
valid today as it ever was.

This book attempts to clarify the issues. Firstly, the basis of existing
classical models is critically examined. Secondly, the data on which the new
stratigraphic framework relies is considered. Then the ways of utilising
this, along with more traditional means, are discussed : the fossil record,
sea level change, glacial and nonglacial environments. Finally, the subject
is placed in perspective overview, and the stratigraphic framework extended by
selective correlation.

ACKNOWLEDGEMENTS

Many colleagues, too numerous to mention individually, discussed points of detail, and I have derived great benefit from personal contact in the Quaternary Research Association and INQUA both at home and abroad. Some, however, have influenced me considerably, both personally and, or, in their writings, and I acknowledge my debt to the following in particular : Dr. G.S. Boulton, Professor K.W. Butzer, Dr. G.R. Coope, Professor R.W. Fairbridge, Dr. G.J. Kukla, Dr. R.B. Morrison and Dr. N.J. Shackleton. Just as my stratigraphic mentor is Professor T. Neville George, F.R.S., to whom I owe much for his friendship and encouragement, my geomorphic mentor is Professor E.H. Brown who, with the late Professors S.W. Wooldridge and S.E. Hollingworth, introduced me to the Quaternary.

At the University College of Wales, Aberystwyth, my grateful thanks are due to many : Professor C. Kidson generously allowed me the full range of his Department's facilities, and Professor H. Carter allowed secretarial time. I am especially grateful to Miss Linda James who typed the text, and to Mr. Michael G. Jones who was responsible for most of the art work and table settings. Miss Jill Anderson, Mr. Tim Cairnes and Mr. Kenn Wass (London) drew the remainder of the diagrams. The photographic work was executed by Mr. David Griffiths, Mr. Howard Williams and Mr. Iain Wright.

My wife, Elizabeth, compiled the references and index, and assisted in a hundred and one different ways : I cannot thank her adequately enough.

Chapter 1
PRELIMINARY CONSIDERATIONS

GENERAL CHARACTERISTICS OF THE QUATERNARY

The Quaternary is a subdivision of geological time which includes the present-
day (Table 1-1). This means that contemporary surroundings are as much a part
of it as the *Ice Age*. Present-day environments, however, are treated largely
from an historical viewpoint, but contemporary processes and spatial distribution studies are used as a basis for inference about the past.

TABLE 1-1 Major units of Standard Global Chronostratigraphic
(Geochronologic) Scale for the Cenozoic Era.

Erathems and Eras	Systems and Periods	Series and Epochs	Isotopic Dating (Ma)*	
			Duration of Unit	Age of Beginning of Unit
	Quaternary	Holocene	0.01	0.01 (10 ka)**
		Pleistocene	ca. 1.6 ka	ca. 1.6 ka
Cenozoic***		Pliocene	5	7
		Miocene	19	26
	Tertiary****	Oligocene	12	38
		Eocene	16	54
		Palaeocene	11	65

*	Millions of years
**	Thousands of years
***	Together with Palaeozoic and Mesozoic the Cenozoic makes up the Phanerozoic Eon
****	Sometimes dispensed with and instead a Neogene System (Miocene and Pliocene Series) and Palaeogene System (Palaeocene, Eocene and Oligocene Series) recognised

Invaluable though this Lyellian maxim is (that, 'the present is the key to the
past'), it is not always easy to put precept into practice because many
Quaternary environments were unique and lack precise modern analogues. Fortunately these difficulties are diminished by well preserved and relatively
abundant evidence, advantages unrivalled by pre-Quaternary data. Such relative perfection makes possible finely detailed reconstructions, and inference
is further assisted because the geographical framework of the Quaternary
remained unchanged: that is, with respect to continental and oceanic configurations, and the fundamentals of atmospheric circulation.

Climatic change is unquestionably the dominating characteristic of the
Quaternary. 'It was originally deduced from signs that certain European glaciers had been at one time more extensive (Table 1-2). Later it was shown that

mid-latitude glaciation on a continental scale had taken place on more than one occasion.

TABLE 1-2   Development of the Glacial Theory*

1795	James Hutton	identifies erratics in the Jura as glacier-borne.
1802	John Playfair	supports Hutton.
1815	Perraudin	a Swiss peasant, infers former Alpine glaciation.
1821	Venetz	communicates Perraudin's ideas, and elaborates on them, to the Society of Natural History, Luzern.
1824	Esmark	recognises former mountain glaciation in Norway.
1829	Venetz	argues that most of Europe was glaciated.
1832	Bernhardi	first recognises continental glaciation in Germany.
1834	de Charpentier	supports Venetz's views.
1837	Agassiz	lectures to the Helvetic Society on 'a great ice period'.
1838	Buckland	is convinced that Britain was glaciated, hence renounces former belief in the biblical flood as professed in his *Reliquiae Diluvianae (1823)*.
1839	Conrad	first acceptance of the theory in America.
1840	Agassiz	publishes *Etudes sur les glaciers*, Neuchâtel.
1840	Agassiz	visits Britain, and is accompanied by Buckland.
1841	de Charpentier	publishes *Essai sur le glaciers et sur le terrain erratique du bassin du Rhône*, Lausanne.
1847	Agassiz	recognises that the Alpine and North European glaciers were separate.
1847	Collomb	recognises 2 glaciations in the Vosges Mountains.
1852	Ramsay	recognises 2 glaciations in North Wales.
1853	Chambers	recognises 2 glaciations in Scotland.
1856	Morlot	recognises 2 glaciations in Switzerland.
1862	Jamieson	publishes 'On the ice-worn rocks of Scotland'.
1863	Archibald Geikie	publishes 'On the phenomenon of the glacial drift of Scotland'.

* see especially North (1943), Flint (1971), Charlesworth (1957) and Chorley *et al*(1964).

It has been estimated that at such times, with the exception of Antarctica, ice sheets and glaciers occupied an area 13 times their present size, and that the average thickness of the larger ice sheets was about 2 km (Flint 1971). In hot deserts lake margins expanded and contracted, though not synchronously between different continents, while more universally, sea level rose and fell through 100 m, a figure which may turn out to be a conservative estimate. Such different alternations were climatically controlled. Differences in mean annual temperature between alternating cold (glacial) and warm (interglacial)

periods were about .5°C in equable maritime locations, but greater than 10°C in some continental interiors. These changes caused a latitudinal migration of the principal geographical zones. During glacial maxima the zones contracted equatorwards, but subsequently expanded to about their present location during times of interglacial warmth. Figure 1-1 shows the contrast between the principal vegetational zones in Europe today, and those which obtained during the maximum of the last glaciation.

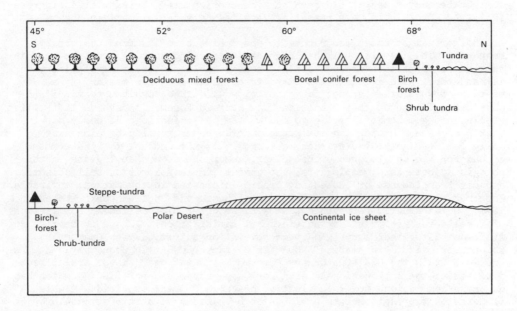

Fig. 1-1.   Interglacial and Glacial vegetation in Europe (after van der Hammen *et al* 1971)

Environmental changes of this magnitude subjected plant and animal communities to repeated stress which caused migration and often extinction. Because of the relatively brief duration of the Quaternary (Table 1-1) only limited tax-onomic and morphological change occured. Thus its palaeontology is concerned more with palaeoecology than phylogeny, with, of course, notable exceptions: e.g. *Homo sapiens* (modern man), appeared some 0.4 Ma ago, having descended from the earlier Quaternary genus *Homo habilis* who migrated from Africa into Asia and Europe. Such exceptions apart, however, morphological similarity between modern and earlier forms allows rather precise palaeoecological infer-ence. This depends not only on an assumption of similar physiology, but also on a sound knowledge of modern ecosystems. Unfortunately comparison between past and present ones frequently reveals how limited knowledge is of the latter.

Two developments in particular since the Second World War have promoted great progress in Quaternary Research. Firstly radiocarbon dating has allowed a remarkably detailed picture of the last glacial episode to emerge; and secondly, exploration of the ocean floors has revealed detailed palaeontol-ogical and geochemical records of climatic change which in some cases are

arguably complete for long periods. This development has opened up novel and
wholly unsuspected means whereby a global standard framework for the correlation of Quaternary events may eventually be achieved. Until this development
most Quaternary research had been continent-orientated, but now it is increasingly the oceans which are revealing detailed data for climatic change.

All historical sciences contribute to and derive benefits from Quaternary
Research. The intellectual challenge of deciphering spatial and temporal circumstances of past environments is so considerable that multidisciplinary work
is mandatory, with a constant need to trade ideas and information. At a
meeting of what other group, for example, would it be possible to find 'geo-
physicists listening to explanations why certain Foraminifera have left-coiled
shells, and geochemists trying to follow the peregrinations of early man in
East Africa, and archaeologists developing an understanding of air-mass
analysis?' (Wright 1973) At such meetings it is not unusual to find common
ground for geologists, geomorphologists, geographers, archaeologists, anthro-
pologists, botanists, zoologists, geochemists, meteorologists, soil scientists
and civil engineers, or indeed any others interested in the natural environment.
Together they provide 'an admirable mix of basic and applied work' (Washburn
1971), which usually represents what H.E. Wright (1973) called 'a combination
of intellectual stimulation and emotional satisfaction ... hard to beat'. So,
it is not altogether surprising that several organisations for Quaternary
Scientists exist: in America AMQUA, in Germany DEUQUA, in Scandinavia NORDQUA,
and in Britain the QRA (Quaternary Research Association). These are all sub-
ordinate to the International Union for Quaternary Research (INQUA) which has
numerous commissions for co-ordinating research in various fields. Table 1-3
illustrates the rapid growth of INQUA as revealed by its International
Congresses every four years.

TABLE 1-3   The growth of INQUA*

Year	Place(s) and Country of Congress	Participants	Countries represented
1928**	Copenhagen, Denmark	102	17
1932***	Leningrad and Moscow, U.S.S.R.	239	9
1936	Vienna, Austria	187	23
1953	Rome and Pisa, Italy	242	26
1957	Madrid and Barcelona, Spain	294	33
1961	Warsaw, Poland	500	31
1965	Boulder and Denver, U.S.A.	900	41
1969	Paris, France	1000	53
1973	Christchurch, New Zealand	450	40
1977	Birmingham, United Kingdom	1000	50

*   Based on Neustadt (1969) with additions.
**   Founding of International Association for the study of the European
       Quaternary.
***   Name changed to International Union for Quaternary Research (INQUA).

QUATERNARY, PLEISTOCENE, AND HOLOCENE

Credit for introducing the term *Quaternary* is due to Desnoyers (1829) who
working in the Paris Basin, applied it to deposits overlying Tertiary strata,
It was defined by Reboul (1833) so as to include deposits with fauna or flora

still living, while Lyell (1839), continuing a palaeontological approach to classification, introduced the term *Pleistocene* (meaning most recent) for deposits which contained more than 70% of mollusca still living. Schimper (1837) had already written of the *Eiszeit* (Ice Age), but it was only after general acceptance of the Glacial Theory (Table 1-2) that Forbes (1846) was able to refine Lyell's definition of Pleistocene further by suggesting that it should be synonomous with the *Glacial Epoch*, with postglacial time being designated *Recent*. Growth of the Glacial Theory was, of course, foundational to much of the interest and progress in understanding the Quaternary during the last century (Table 1-2).

Pleistocene and *Glacial Epoch* should, however, no longer be regarded as the synonyms that Forbes intended them to be, for glaciation is known to have occured before the opening of the Quaternary. Nowadays it is more accurate to refer to *The Late Cenozoic Ice Ages* (e.g. Turekian 1971). Again, as things turned out, the classical ice age of Europe only occupied the last 850 ka of the Pleistocene; hence usage of the term 'pre-glacial Pleistocene' (e.g. Shackleton and Opdyke 1973).

The geological status accorded to the Quaternary has varied. Charlesworth (1957) for example, referred to it as an Era, thus logically continuing the classification of Primary (Palaeozoic), Secondary (Mesozoic) and Tertiary. *Tertiary* and *Quaternary* are in a sense anachronistic because Primary and Secondary are no longer used: instead both are incorporated within the Cenozoic Era. Indeed Flint (1957) urged the abandonment of both, yet in the successor to his 1957 work he introduced *Quaternary* into the title so as to 'facilitate understanding of the subject matter' (Flint 1971). By now the term is so firmly entrenched that it is difficult to see what advantage accrues were it to be abandoned, not that the International Geological Congress would ever be likely to accept any such proposal.

Although Forbes proposed the term *Recent* in 1846, it was not until 1885 that the International Geological Congress accepted a term for that period, when *Holocene* (meaning wholly recent) was established. Even now some believe that any subdivision of post-Pliocene time by recognising more than a Pleistocene Series is unecessary (Flint 1971. West 1968). But others have argued persuasively for the retention of a Holocene (or Recent) Series (e.g. Morrison *et al* 1957). And Fairbridge (1968) declared firmly that all stratigraphic nomenclature ought to be based on its usefulness in promoting understanding and communication; moreover, the nearer to the present-day, the shorter the subdivision ought to be. It is to be hoped that recognition of both a Pleistocene and Holocene Series by the International Subcommission on Stratigraphical Classification will now settle the issue (Hedberg 1976).

OBJECTIVES

For the geologist the scientific goals are clear enough because the Quaternary is simply the latest chapter of earth history. Yet it is sufficiently different from pre-Quaternary time to create some problems of approach. To some extent these may be illusory, and arise from the greater degree of resolving power available. Real or apparent, however, they are considered later. In no way is the geologist's fundamental goal changed, namely to determine the fullest possible statement about the earth's history during the period concerned.

As noted earlier many others have an historical stake in the Quaternary.
Development of particular phenomena through time is a principal interest, such
as vegetational or hominid history. Another is to provide an explanation for
contemporary environmental frameworks, such as landforms, or for distributions,
such as those of rare arctic-alpine plants. It is immaterial whether these
are the result of past or present processes, although usually they result from
interaction of both, thus 'while "everybody" is uniformitarian in accepting
as method that the present is a key to the past, it may no less be true,
history the mentor, that the past is a key to the present' (George 1976).

In a rather more utilitarian role Wright (1973) assembles cogent examples how
Quaternary research may provide answers to contemporary problems. He cites
the heavily polluted lake which it is desired to restore to its natural condition. Yet few historical records exist to demonstrate what such a state was
like. The answer comes from a palaeoecological investigation of fossil assemblages in recent, pre-man, lake sediments, thus establishing a base-line.
Evidence of former processes should also allow prediction of side-effects
resulting from diverting a river channel, or constructing a dam. Even the
civil engineer benefits from knowing not only the origin of hill-slope sediments, but also their age, as exemplified by work on the stability of
natural slopes in south east England (Weeks 1969). More surprisingly perhaps
an ecological role for fire in natural forest ecosystems in North America has
been clearly demonstrated by historical data, and in implication for present
management points to controlled burning as beneficial, a conclusion in utter
contrast to a more traditional fire-protection philosophy (e.g. Heinselman
1973).

Understanding environments past and present are objectives which are clear
enough, but a fully-fledged science ought to possess predictive capabilities
even though much of Quaternary research falls within the ambit of what
Pantin (1968) called the unrestricted sciences: that is, geology and biology,
as opposed to physics and chemistry. Pre-eminent among any such predictive
capabilities for Quaternary research would be a theory of climatic change.

Up until 1960 some 53 theories of climatic change had appeared (Schwarzbach
1960). Most were concerned with past events, and advances in knowledge have
compelled a majority of them to fall by the wayside. Presently, however,
there are signs that significant progress will not be delayed for much longer.
Of the three great realms of planet earth: lithosphere, biosphere and
atmosphere, Flint (1974) has graphically described how a theory of the dynamics of the biosphere grew as a nineteenth century synthesis of the way in
which species evolve. And, how technological advances of the mid-twentieth
century allowed exploration of the oceans leading to a theory of the lithosphere (plate-tectonics). A theory of the atmosphere with power of prognos-
tication has yet to emerge, but Flint believes that with a continuation of
present trends it may be a reality by the end of the century.

Implicit in Flint's belief is the corollary that it will come about, just as
theories of the lithosphere and biosphere did, by the understanding of a past
process and its ongoing continuation. Any realisation of such an objective
will have come about largely through the efforts of a group of scientists who
comprise CLIMAP (Climate, Long Range Investigation, Mapping and Prediction),
(Hays and Moore 1973), whose object is to monitor climatic change in ocean
cores throughout the world and to map the resultant patterns.

METHOD

Different disciplines pursuing a variety of interests find common ground in
the establishment of a sequence of Quaternary events. Method should, therefore, be self‑evident. Unfortunately while everybody is familiar with basic
stratigraphical principles these are not always followed for a variety of
reasons. Difficulties arise from unusual stratigraphic relationships. In
valley systems, for example, younger deposits frequently, though not invariably, lie altitudinally below older ones and may be separated by a considerable gap. Others are patchy and discontinuous so that succession is not easy
to establish and correlation is fraught with the risk of considerable error.
Because of the short duration of the Quaternary these difficulties cannot be
resolved by classical palaeontological methods based on the evolution of
species, because significant development is only found in a few groups. Hence
the limited use of a method principally successful in the subdivision and
correlation of pre-Quaternary Phanerozoic rocks. This situation is further
impaired because the majority of Quaternary deposits are continental in origin
and largely unfossiliferous. Even the most powerful available correlation
tool, that of radiometric dating, depends on particular materials for assay,
and assumption that samples have remained as closed systems since events which
it is sought to date. Thus inter-regional correlation is difficult, and inter-
continental correlation beset by many problems, a situation not made easier by
a tendency for some disciplines to regard their method as the most appropriate.

Using the metaphor of a jig-saw, pieces of Quaternary data require description
and interpretation in precise and unambiguous terms, so as to be readily understood by a wide variety of scientists of different nationalities, before they
may be fitted together at any temporal level. Unfortunately these objectives
have not been realised frequently. Indeed it could be said that force-fitting
of the pieces into preconceived pigeon-holed classifications is what is almost
a way of life for the Quaternary worker. Fortunately geological procedure,
despite any difficulties peculiar to the Quaternary, is clear : (1) local
identification and description of rock units in sequence, (2) correlation of
local sequences, (3) interpretation of the stratigraphic record in terms of
earth history. The third stage is basically an exercise in classification of
the chronology of events, and it is the one which has given rise to the majority of problems.

Despite the logical procedure outlined above, it is not always followed. Many
continue in apparent ignorance of it. Others, though soundly based stratigraphically, do less than justice to their results because of unclear and ambiguous presentation. Hence the desirability for uniformity, not necessarily
of approach, but at least of methods and terminology of presentation. From
such a need has arisen several recommendations, or codes of practice, applicable to the rocks of the Quaternary as well as of earlier Systems.

---

*   Unless specified otherwise, e.g. lithostratigraphical correlation,
     the term *correlation* in this book is taken to mean *time-equivalence*.

The absence of a universally accepted set of rules of practice, a constanly
changing terminology, a seemingly unintegrated and contradictory mass of data,
and a number of frameworks for classifying events, makes it unsurprising that
the student on being introduced to the Quaternary finds it complex and
bewildering. Any sign of improvement is slow to appear: at the 1957 INQUA
Congress in Madrid, van der Vlerk reported that a request for a stratigraphical table of the Quaternary sent to 22 countries had produced 22 different
answers. At the 1973 meeting the president lamented woefully - ' I wish I
could say that the position was substantially better today' (Mitchell 1973).

In these circumstances it is remarkable that some classifications have lasted
so long and their terms continue to be used (Table 1-4). Of these the most
venerable is the Alpine model of Penck and Bruckner (1909).

TABLE 1-4 Some Classical Schemes of Subdivision*

Alps	Northern Europe	British Isles	North America	East Africa**
WÜRM	WEICHSEL	NEWER DRIFT	WISCONSIN	MAKALIAN GAMBLIAN
R-W	Eemian	Ipswichian	Sangamon	inter-pluvial
RISS	SAALE	GIPPING	ILLINOIAN	KANJERAN
M-R	Holstein	Hoxnian	Yarmouth	
MINDEL	ELSTER	LOWESTOFT	KANSAN	KAMASIAN
G-M	Cromerian	Cromerian	Aftonian	inter-pluvial
GÜNZ			NEBRASKAN	KAGUERAN

*   GLACIALS and *interglacials*
** PLUVIALS and *interpluvials* (Woldstedt 1965)

The terms Günz, Mindel, Riss and Würm, however, have yet to be defined stratigraphically so that the model can hardly be regarded as adequate. Indeed the
events which these terms refer to (even with the addition of the previous
Donau I, Donau II 'Biber' glaciations) do not represent all of Quaternary time.
Even the time-span approximately covered, some 850 ka, is infinitely more
complicated than is suggested by four simple glaciations and intervening interglacials. Despite this the terms are still used and events recognised elsewhere correlated with the Alpine model.

Tendencies to oversimplify in this way lead to new discoveries being forced
into a pigeon-holed classification. Such arbitrary methods tend to perpet-
uate an illusion of security and precision in an apparently repeated confirmation of the original model. This tendency to confirm discoveries from
limited amounts of data has been called *The Reinforcement Syndrome* by
Watkins (1971), who cites the classical four-fold Alpine subdivision of the
Pleistocene as an example. As a result of it labels are used in areas remote
from the original localities, and correlation proceeds through completion of
conceptual classification boxes. This unfortunate tendency was strengthened
by relating the Alpine chronology to Milankovitch's climatic chronology, which
was based on mathematical calculations of secular variations in the eccentri-

city of the earth's orbit (periodicity ca. 90 ka), tilt of its axis (periodicity ca. 40 ka), and precession of the equinoxes (periodicity ca. 20 ka),
(e.g. Zeuner 1959). It led to what Kukla (1975) called a 'count from the top'
method of dating: that is, from Würm downwards, or backwards, in time. In
this way a fossil soil in Austria, the Kremser palaeosol, was estimated to be
of Riss-Würm interglacial age; whereas later work showed it to be about a
million years older. Similar 'count from the top' procedures are still widespread and lead to intellectually unexacting chronological classifications.

Even though the last glaciation is now well known in outline a persistent
usage of inadequate models frequently causes confusion. Use of the term Würm
in Europe, for example, is still not free from equivocation, much as it was
when Büdel (1949) referred to the *Würm-wirwarr* (Würm-mess). Because events
preceding the last glaciation are less well understood it will be instructive to consider the basis of existing models of classification before proceeding further.

Chapter 2
CLASSICAL MODELS

Long-established classifications of the Pleistocene have been, and still are,
used in: the Alpine regions of Germany, Austria, France, Italy and Jugoslavia;
Northern Europe, including both German republics, Belgium, The Netherlands,
Denmark, Norway and Sweden; The United States and Canada; East Africa and
other parts of that continent; and the British Isles. In their essentials
all were formulated before the recent wave of research on deep sea cores which
have produced a potentially complete stratigraphic record for the Pleistocene
(Chapter 3). In terms of their individual antiquity, therefore, all of them
are deficient to some degree. In some of them method is equally deficient.
One or two have common characteristics, but all have serious drawbacks as will
become apparent when they are now examined in turn.

THE ALPINE MODEL

The celebrated Alpine scheme of glaciations, which has influenced so many
others, was devised largely by Albrecht Penck (1885) between the Iller and
Lech rivers in the Bavarian plateau country south of Ulm, Augsburg and Munich
(Fig. 2-1). It was subsequently elaborated and extended, firstly to the
Alpine region in general and secondly farther afield. Even today the terms
employed by this classification are used confidently in continents remote
from Central Europe.

In its essentials the scheme involves recognition of four glaciations from
four discrete fluvioglacial outwash terraces. While it is based partly on
stratigraphy, it rests almost entirely on morphology - *a morphostratigraphical**
basis which should not be obscured by either the various refinements which
followed, nor by Penck's claim that each glaciation is represented by a 'series'
of glacial deposits. Its principles are straightforward: four outwash
terraces (*schotter*) were mapped (morphological mapping being the principal
procedure) and traced up-valley to where they merged with end-moraines marking
the extent of contemporaneous glaciers (Fig. 2-2). Originally only the two
lowest terraces were linked with end-moraines, but subsequently all four were
so related. Figure 2-2 allows the principles to be explained : the first
glaciation is indicated by the outer end-moraine, downstream from which the
upper outwash plain (*schotter*) formerly filled the valley floor. Subsequently,
after deglaciation, the outwash was eroded and valley deepening dissected it
into a terrace. Enough of the terrace survives to show its actual or former
connection to the end-moraine lying upstream. Such downcutting and dissection
was deemed to have occurred during an interglacial interval. Note, in passing,
therefore, that interglacial time is not here represented by any deposits;
but instead by an erosional episode. The entire process described was then
repeated during a subsequent glaciation. This, however, as the diagram shows
(Fig. 2-2), was not as extensive as the earlier glaciation and its end-moraine

---

*A *morphostratigraphical unit* is a body of sediment classified principally on
surface form (Chapter 4).

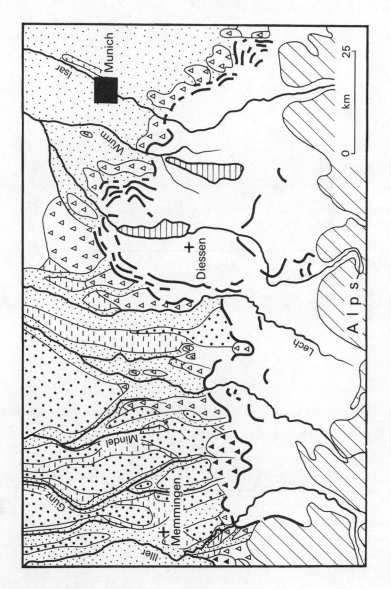

Fig. 2-1. The Bavarian Foreland (after Penck and Bruckner 1909). Heavy lines : moraines. Open triangles : Riss moraine. Closed triangles : Mindel moraines. Heavy dots : deckenschotter. Dash : high terrace. Fine dots : low terrace

lies inside the outer one. Its contemporary outwash was transmitted downstream and deposited altitudinally below, or against the dissected margins, or
sometimes partly overlaid, the earlier one. Subsequent interglacial erosion
and dissection of the newer body formed the lower terrace. Like the earlier
one it can be traced upstream to its contemporaneous end-moraine where it
interdigitates with morainic deposits (Fig. 2-2).

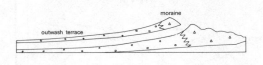

Fig. 2-2.   Outwash terraces (schotter) and end-moraines

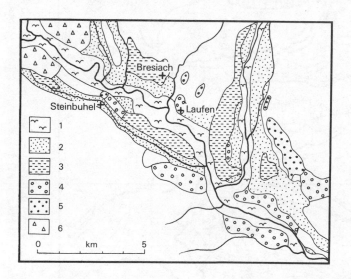

Fig. 2-3.   Terraces in the Save Valley (after Penck and
Bruckner 1909). Key: 1 : alluvium. 2 : Low terrace. 3 :
High terrace. 4 : Younger Deckenschotter. 5 : Older
Deckenschotter. 6 : Glacial drift

In Albrecht Penck's (1885) original scheme he only recognised three glaciations. These were based on three schotter: the Deckenschotter, Hochterrassen and Niedterrassen, in Bavaria and around Lake Constance. Soon afterwards Eduard Bruckner (1886) recognised them in the Salzach valley farther east. Then in 1909 both of them published their classic work: *Die Alpen im Eiszeitalter* (Penck and Bruckner 1909) which subsequently became so influential throughout the world. In it Penck subdivided his Deckenschotter into Older and Younger units, thus allowing recognition of four glaciations. These were named after four Bavarian rivers (Table 2-1):

TABLE 2-1 Penck and Bruckner's scheme of glaciations (with related terraces) and interglacials in the Alpine region

	erosion
Würm Glaciation	Niedterrassen (Low Terrace)
Riss-Würm interglacial	*erosion*
Riss Glaciation	Hochterrassen (High Terrace)
Mindel-Riss interglacial	*erosion*
Mindel Glaciation	Younger Deckenschotter
Günz-Mindel interglacial	*erosion*
Günz Glaciation	Older Deckenschotter

Both the High and Low Terraces were readily linked with two belts of prominent end-moraines: the Low Terrace with the inner moraines (*Jung-Endmoraine* on Figure 2-1), which are freshly defined as morphological features and little modified by post-formational processes; and the High Terrace with the younger part of the outer moraines which, in contrast to the younger ones, are morphologically subdued. Further points of contrast are in the drainage pattern which is said to be more 'mature' on the older deposits; whereas on the younger ones it is less organised and accompanied by numerous small lakes. Furthermore, the old moraines, unlike the younger ones, are covered by loess.

Figure 2-3 is Penck and Bruckner's map of part of the Save valley and shows clear separation of the terrace fragments (Fig. 2-4) due to the considerable amount of 'interglacial' erosion which took place. In other regions where less erosion occurred, such as near Munich, the schotters are superimposed. It is only where deep erosion obtained, often in response to presumed uplift, that clear separation of the schotters is found.

The High and Low Terrace both lie within the valley (Figure 2-3), as they do in all the principal ones throughout the Alpine region. When extensive preservation occurs it is possible to reconstruct their long profiles on a heigh-distance diagram (Fig. 2-5). The Deckenschotter are invariably greatly

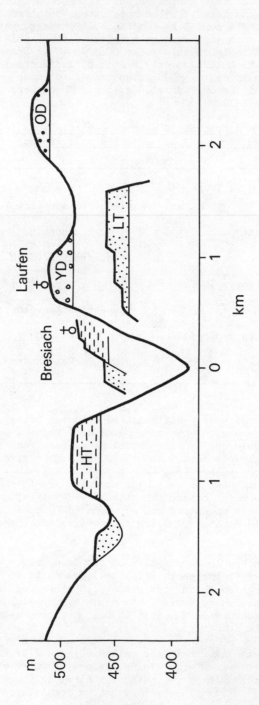

Fig. 2-4. Cross-section of Save Valley near Laufen (after Penck and Bruckner 1909). OD : Older Deckenschotter. YD : Younger Deckenschotter. HT : High Terrace. LT : Low Terrace

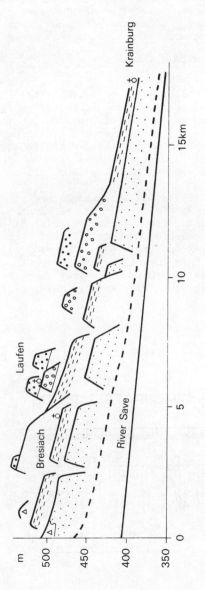

Fig. 2-5. Long profile of the Save River upstream from Krainburg with principal outwash terrace fragments plotted. The broken line indicates the extent to which the river has down-cut into solid rock. Open triangles = ground moraine of ice advances contemporaneous with outwash deposition (after Penck and Bruckner 1909)

dissected (Figs. 2-4 and 2-5). In Bavaria the Older Deckenschotter consists of fragmentary sheets of gravels extending from the foot of the Alps. A considerable variation in the altitude of remaining fragments was accounted for by differential uplift since their formation. Surviving gravel sheets *decken*, survive on isolated plateau and hills (frequently bearing churches or settlements), and which serve not only to demonstrate considerable erosion and dissection but also that the details of their contemporary geomorphology is lost.

Supporting data

Near Deisenhofen, in the Gleissen valley south of Munich (Fig. 2-1), Penck described three superimposed gravel units: 'Deckenschotter', High Terrace, and Low Terrace gravels. All three are weathered, but the Deckenschotter is more so than the others (Fig. 2-6). This is characteristic of both Older and Younger Deckenschotter throughout the Alpine region, with limestone constituents having been dissolved and crystalline clasts thoroughly rotted, and is represented by a strongly weathered red clay zone called *Ferretto*.

Organic beds show that complete deglaciation occurred between the glaciations. These include: (1) the celebrated *Hottinger Breccia* near Innsbruck, a colluvial and scree deposit containing flora of interglacial aspect; (2) various *schieferkohlen*, which are heavily compressed layers of peat (lignite) containing plant remains; (3) lake deposits; (4) calcareous tufas; and (5) other plant bearing deposits. Nearing all of these were thought to belong to the Riss-Würm interglacial, deposits overlying them are almost invariably those of the Würm glaciation. W.B. Wright (1914. 1937) has

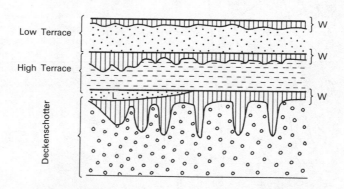

Fig. 2-6   Weathered gravels near Deisenhofen (after
            Penck and Bruckner 1909). W : Weathering
            horizon. L : Loess.

provided a good English summary of Penck and Bruckner's (1909) principal
conclusions drawn from such data.

Duration of glacials and interglacials

Penck and Bruckner (1909) also provided an estimate of the relative duration
of the four glaciations, and of the interglacials. The Riss glaciation was
thought to have been longer than the Würm because a greater quantity of
glacial and fluvioglacial deposits accumulated at that time. By similar
reasoning, the Günz glaciation was deemed to have lasted the longest. They
attached more reliability to their estimates for interglacial time, and in
particular thought that the Mindel-Riss interglacial lasted much longer than
the others. It seemed to them that more geological work was accomplished
at that time than at any other: terraces of the Günz and Mindel lie high
above those of the Riss and Würm glaciations; Günz and Mindel deposits are
greatly weathered in comparison with younger ones; Mindel moraines are
greatly denuded; and whenever they were able to identify interglacial
deposits of that period they tended to be thicker than those of of Riss-Würm.
Their estimate for the relative duration of each interglacial is:

Günz-Mindel	Mindel-Riss	Riss-Würm	post-Würm
3	12	3	1

From this arose the concept of a *Great Interglacial*, a notion universally
extended by Zeuner (1945. 1959). As will be seen later (Chapter 3) there is
no basis for it, but regretably, it continues to obtain currency.

Elaboration and Assessment

Such is the basis of the 'Alpine' model. When Table 2-1 is examined carefully
it is apparent that Pleistocene time, as conceived by Penck and Bruckner
(and indeed by everyone else who follows their scheme) is represented in the
type area by glacial and fluvioglacial (outwash) deposits that only represent
part of that time. A good deal of Pleistocene time is represented by
erosional breaks in the stratigraphic record. Moreover, from what is now
known about glacial pulse during the Pleistocene it could be argued that each
end-moraine and its associated outwash terrace only represents a few thousand
years at most of one cold stage or glacial cycle. Strictly on the basis of
the model as presented thus far, interglacial time is not represented by
sediments, but merely by inferred erosion during which outwash schotter were
dissected into terraces. In other words not only do the sediments not
represent the full time-span under discussion, but the unconformities between
each successive terrace probably conceal events lost to the record either
locally or regionally. As such, then, because it is primarily a morphostratigraphical model, it is inherently deficient.

The notion that some of the Alpine gravel deposits might be interglacial
rather than glacial in age antedates Penck and Bruckner's (1909) work. But
it was not until the early years of the 1920's that Penck fully accepted and
elaborated upon such suggestions. Modern work has amply demonstrated this.
Mollusca cf interglacial aspect have been found in the High Terrace gravels
near Munich (Brunnacker and Brunnacker 1962), and at Moosburg - the *Moosburg
Gravels* (Nathan 1953). Radiocarbon dating showed that some Low Terrace gravels
are Holocene in age (Gaul 1973). And Brunnacker (1962) has described inter-

glacial soil horizons in the Low Terrace at Hörmating.

Major modification of the four-fold classification, however, had to await
Eberl's (1930) recognition that the four glaciations were compound; and
moreover, there was evidence for a pre-Günz advance which he named the Donau,
though its terraces were unconnected to any moraine-ridges. His work
recognised: (1) 5 pre-Günz terraces, the Ottobeuren gravels which were
fluviatile, the Staufenberg gravels not certainly of glacial origin, and
three sub-stages of the Donau glaciation; (2) 2 Günz sub-stages; (3) 2
Mindel sub-stages; (4) 2 Riss sub-stages; and (5) 3 Würm sub-stages. Like
the earlier work, all of these were recognised on a combination of morphological mapping and lithostratigraphy of gravel terrace units. In the case of
the Günz Deckenschotter he argued that Penck and Bruckner's (1909) notion of
tectonic displacement was incorrect, and instead that terrace remnants at
different heights were different sub-stages: i.e. Günz 1 and 2, and Donau 1
2 and 3. In passing it must be noted that Eberl (1930) was heavily
influenced by the ideas of Milankovitch.

Further elaboration was made by Schaefer (1953) who recognised a glaciation
earlier than the Donau which he named the Biber advance. Like Eberl's
Donau glaciation it was recognised from a gravel terrace unit unconnected to
a moraine-ridge. More significantly, however, Schaefer (1953) showed that
the episodes of erosion which took place between successive terraces were
glacial rather than interglacial as proposed by Penck and Bruckner (1909).

Although some of Eberl's work has been shown to be invalid, most of it still
stands. What is unclear is the nature of the intervals between his various
sub-stages, a situation made difficult because the separation of sterile
glacial and interglacial gravels is impossible. Palaeobotanical investigations,
long hailed as eminently successful in Northern Europe, have not resolved
the dilemma. Frenzel's (1973) assessment of the available data led him to
the conclusions summarized on Table 2.2. If, as he argued, the Zeifen
interglacial (Chapter 6) corresponds in part to the Eemian of Northern
Europe, then both Riss and Wurm glaciations in the Alps, separated by the
Furamoos interglacial (Chapter 6), postdate what is conventionally thought
of as being the 'Last Interglacial' by most European workers. But note that
the exact status of Eemian is unclear (below). His Pfefferbicl interglacial
may possibly correlate with the Holstein of Northern Europe, but the age of
the Uhlenberg interglacial is unknown. Note also, that the *schieferkohlen*
are now thought to be of different interglacial ages, or even interstadial
age shown by radiocarbon dates of 40 to 30 ka BP (Frenzel 1973).

Conclusion

The Alpine classification is clearly unsatisfactory. Yet despite this the
terms Günz, Mundel, Riss and Würm continue to echo confusingly throughout the
world. It would seem that a conscious effort is required to abandon them as
labels referring to intervals of time - even in the Alpine region itself.
Kukla (1975), for example, advocates their usage only in connection with
specific terraces of specific moraine-ridges in the Alps: e.g. Würm terrace
or Würm moraine. That is, as morphostratigraphical terms and not as time
terms (chronostratigraphic or time stratigraphic units). Great care must be.
taken so as to avoid confusing what are terms relating to morphostratigraphy
and terms relating to time. By way of example he writes (Kukla 1975):

TABLE 2-2 Possible palaeobotanical correlations by
           Frenzel (1973)

| Conventional correlation | | Frenzel 1973 |
Alps	Northern Europe	Alps
Würm	Weichsel	Würm
		Furamoos
		Riss (whole or part)
R-W	*Eemian*	*Zeifen*
Riss	Saale	
M-R	*Holstein*	? *Pfefferbichl*
Mindel	Elster	
		? *Uhlenberg*
G-M	*Cromerian*	
		Donau (whole or part)

the gravels of the lower terrace (Niederterrace) close to
Ostrava (north central Czechoslovakia), were for many
years mapped as being of Würm age, which they really are.
But after the famous Czech Quaternary stratigrapher.
Tyracek, unearthed a rusty bicycle steering rod from the
intact gravel, this level was recharted as Holocene
alluvium, which it is also. In this manner a shocking
conflict with established stratigraphic schemes was
avoided. But the logical conclusion escaped: the gravels
of the Niederterrasse are by definition Würm gravels,
whether they contain bicycles or Roman bricks, because
the Würm gravel terrace is a morphostratigraphic, not a
time stratigraphic unit.

While 'Würm' terraces span more than one glacial advance, as well as the
intervening non-glacial intervals - which include part of the postglacial
(Holocene), the actual Würm moraine-ridges and immediately adjacent outwash
terraces merely represent a few thousand years at the most. 'Würm' as a
morphostratigraphic concept is clearly not synonomous with 'Würm' as a time
(chronostratigraphic) concept. Furthermore as it is known that the time-span
covered by the terms by Günz, Mindel, Riss and Würm witnessed ten major global
glaciations (Chapter 3), then the sooner the better their demise as major
universal time artefacts for the Quaternary occurs.

NORTHERN EUROPE

The classical sequence of glacials and interglacials in Northern Europe
(Table 2-3) is based largely on a subdivision of the deposits of the
Scandinavian Ice Sheet. Classification rests almost exclusively on the
recognition that major end-moraine systems, which trend across the North
German Plain and its extension (Fig. 2-7), effectively delimit successive
glaciations. These, named after various rivers, are the Elster, Saale,
Warthe and Weichsel Glaciations. All but Warthe were named by Keihlack
(1926). End moraines of the Warthe and Weichsel Glaciations are fresh in
appearance, and still retain to a large extent their original morphology,
as well as bearing numerous closed depressions on their surfaces. In
contrast, those of the Saale Glaciation, while still retaining their original
morphology, are flatter, more subdued, and have been dissected to a greater
degree. End-moraines of the Elster Glaciation have been denuded to such an
extent that they have lost their original morphology. (Woldstedt 1953).

TABLE 2-3 The classical sequence of glaciations and
           integlacials in Northern Europe

Wiechselian Glaciation
Eemian Interglacial
Warthe Glaciation)   ) Saalian Glaciation Drenthe Glaciation)
Holstein Interglacial
Elster Glaciation
Cromerian 'complex' of interglacials

In its reliance on landforms there are clear comparisons between this and the
Alpine scheme: both are morphostratigraphically based. Yet, unlike the
Alpine region, subdivision is confirmed and amplified by numerous interglacial deposits of marine and continental origin, whose stratigraphical
position is established with reference to both overlying and underlying
glacial deposits. Such apparent precision, however, should not obscure the
fundamentally morphostratigraphic character of the classification.

Elster Glaciation

Elster glacial deposits are well known in the Hamburg area, where locally
they reach up to 70 m in thickness. In most areas, however, they are buried

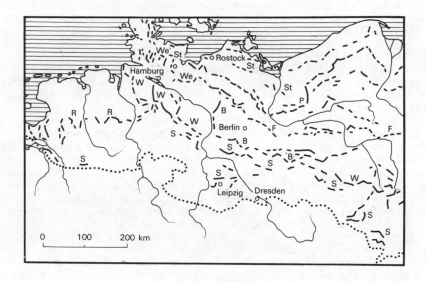

Fig. 2-7.   End moraines of the North German Plain (after
Woldstedt 1967).   S : Saale. R : Rehburger. W : Warthe. B :
Brandenburg. F : Frankfurt. We : Weichsel. P : Pomeranian.
St : Stettin. Maximum glaciation : dotted line

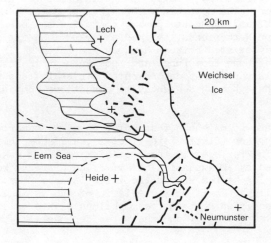

Fig. 2-8.   Warthe moraines breached by
Eemian deposits (after Woldstedt 1954)

by, and difficult to distinguish from, the subsequent Saale glacial deposits.
Other than in Thuringia and Upper Saxony they do not outcrop at the surface.
As befits the oldest glacial deposits in the sequence they are character-
istically leached and greatly oxidised, and on occasion dissected to such an
extent that all that survives is a lag deposit of erratic boulders.

Stratigraphicaly they are younger than the 'Cromerian Complex' of interglacials (Chapter 6). In north west Germany the *Lauenburg clay* is a
widespread marker horizon consisting of sandy clays and clays. It overlies
the Elster till, and is believed to be a meltwater deposit formed during and
immediately after deglaciation.

It has not proved possible to demonstrate beyond any doubt that the
Netherlands was glaciated at this time. Unequivocal till deposits do not
exist, or remain to be discovered. There are indications that the ice front
was no great distance away, for the problematical 'poklei' (pottery clay)
deposits may have been deposited in valley systems draining to a large ice-
dammed lake in the North Sea (Zonneveld 1958). A further problem associated
with them is that they lie in deep and narrow basins whose origin is not
understood. Stratigraphically they are pre-Holstein in age, and some have
argued that they are the same age as the *Launenburg clay* of north west
Germany but this is by no means certain (de Jong 1967).

Ice from eastern and western Scandinavia effected glaciation : in Lower
Saxony high percentages of erratics from eastern Scandinavia occur (Luttig
1960), but not around Hamburg (Richter 1962). Richter proposed five
advances, separated by interstadials when the ice withdrew, but this rests
entirely on lithological contrasts between different Elster tills.

Holstein Interglacial

Deposits of the Holstein Interglacial overlie those of the Elster Glaciation
and are, in turn, overlain by, caught up and deformed in, glacial drift of
the Saale Glaciation. The Holstein marine transgression penetrated from the
north or north-west, and the contemporary coastline ran more or less west to
east in the northern Netherlands, defined a marked indentation along the
Elbe, while a seaway existed between Hamburg and Lübeck. In the Baltic
marine deposits are known from Fünen and Mecklenburg. Palaeoecological
inference from molluscan faunas demonstrated a rise in temperature, initially
to boreal conditions, then subsequently to conditions comparable to the
present-day North Sea.

The vegetational succession shows that after an early subarctic climate with
birch pine and *Hippophae* (Elster Late-glacial), a birch-pine episode
followed before being replaced by the expansion of Mixed Oak Forest. Later
this was accompanied by an expansion of *Abies* and *Carpinus* before the
climatic deterioration at the end of the interglacial. *Picea* is typical of
the first half, while *Abies* is characteristic of the second half of the
interglacial. In addition certain thermophilous plants existed : *Vitis
silvestris, Buxus sempervivens*, and the water fern *Azolla filiculoides*.

Saalian Glaciation

End-moraines of Saalian age, which are related to various 'Middle Terraces'
in the northward flowing German rivers, are morphologically subdued, though
still clearly recognisable as such. This however, refers more to those of

the Drenthe Glaciation (Table 2-3), for the Warthe end-moraines are still
freshly defined. It has long been known that the Saalian Glaciation was
complex, and the number of possible oscillations of the ice margin is still
under active discussion. In West Germany, however, opinion is largely
unanimous in agreeing that the Saalian was not punctuated by an interval of
interglacial status (Duphorn *et al* 1973). The precise nature of the interval
between Drenthe and Warthe Glaciations, however, remains problematical. A
variety of deposits formerly thought to date from between the two advances
have since been assigned to other periods : the diatomite of the 'Ohe warm
phase' seems to be related vegetationally to the Holstein Interglacial; and
the interglacial deposit of the Gerdau valley is now thought to be Eemian.
A 'Treene period', represented by palaeosols, has yet to be confirmed as
intermediate between Drenthe and Warthe. All this, however, does not
discount the notion that the Saalian was subdivided into several distinct
pulses. Unfortunately most proposals of this nature are based on contrasts
in till lithology, morphostratigraphical evidence of end-moraines, and an
interpretation of glacio-tectonic structures in certain moraine ridges.

Luttig (1954. 1968) suggested that the Drenthe Glaciation consisted of
advances ('Staffelin') separated by periods of withdrawal ('Eteppen')
(Fig. 2-9). His scheme is not without complications, for in addition to
relying on lithological differences in tills, it involves controversial
interpretation of the Rehburg Phase and its end-moraines. According to
Luttig maximum glaciation occurred during the Hameln phase (Fig. 2-9) : in
southern Lower Saxony its deposits overlie Middle Terrace deposits of the
Weser, Leine and other rivers. But the Rehburg phase was thought to
antedate this - so that its end-moraines were necessarily overridden during
the Hameln phase. Moraine ridges of the Rehburg phase are topographically
distinctive (that of the Eastern Veluwe in the Netherlands is over 100 m
high), and are characterised by glacio-tectonic imbricated sets of structures.
The tectonics are interpreted by Luttig and others as the result of a
subsequent ice advance over the moraines (glacial overridding). Zagwijn
(1974) believes that ice-surging may be the cause of such renewed forward
movement of the ice-margin. On the other hand, Woldstedt (1954) thought the
Rehburg phase was a simple advance of the ice-margin, a view also reached by
Richter (1961) who believed the Reburg moraines to be younger than the
maximum expansion of the Saalian Ice Sheet. Due to the discovery of 'ground-
moraine' on the ridge crests, current opinion in West Germany regards the
moraines as having been overridden (Duphorn et al 1973).

Much debate has obtained on the age of the freshly defined hummocky end-
moraines of the Warthe Glaciation. At one time Zeuner (1959) argued that the
glaciation was Early Weichselian in age on the basis of the freshness of its
topography. But Woldstedt (1954) had shown in southern Holstein that marine
interglacial deposits of the Eemian transgression actually overlay Warthe
deposits; moreover, that the transgression breached the Warthe end-moraine
arc (Fig. 2-8). On such evidence the case for a pre-Eemian age of the
Warthe Glaciation appeared firmly based. Several Warthe phases seem to have
occurred, although these are based on the evidence of end-moraines, some of
which are deemed to have been overridden, and lithological contrasts between
tills. The precise nature of the interval between Drenthe and Warthe
Glaciations remains obscure.

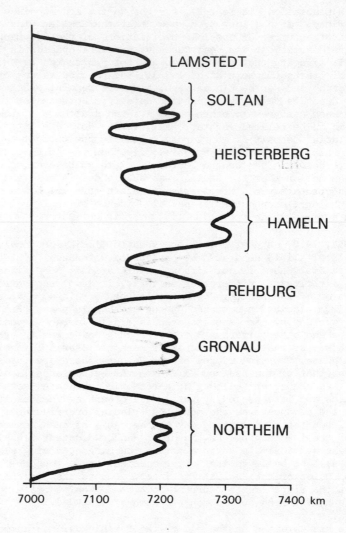

Fig. 2-9.   Phases of the Drenthe Glaciation according
to Luttig (1968). Maximum glaciation is marked by the
Hameln Phase. The Rehburg Phase consists of overridden
moraines. Distances from ice sheet centre. Time
(vertical) not to scale.

Eemian Interglacial

Named after the stream of that name in the eastern Netherlands, marine
interglacial deposits of the Eemian Interglacial are clearly younger than
the Warthe Glaciation (above), and in northern Germany are overlain by
glacial beds of the Weichselian Glaciation. Eemian deposits extend from the
Netherlands across north Germany to the European U.S.S.R. Their fauna is
characterised by *Lusitanian* elements, including *Tapes aurens* var. *eemiensis,*
which demonstrate a warmer sea than the present day.

The vegetational succession of this interglacial is well established, and it
is sufficiently distinctive to be clearly separated from the earlier Holstein
Interglacial. Table 2-4 outlines the principal pollen zone proposed by
Selle (1962) and the climatic inference made from them.

TABLE 2-4   Pollen zones and inferred climatic changes of
            the Eemian Interglacial (after Selle 1962)

	Pollen Zone	Climate
VIb	*Pinus*	Subarctic
VIa	*Pinus-Picea*	
Vb	*Pinus-Picea-Abies*	Cool
Va	*Picea*	
IVb	*Carpinus-Picea*	Temperate
IVa	*Carpinus*	
IIIc	*Corylus-Tilia*	Warm. Climatic
IIIb	*Corylus-Quercus*	Optimium
IIIa	*Corylus-Quercus-Pinus*	
IIb	*Pinus-Quercus*	Warm. Continental
IIa	*Pinus-Betula*	
I	*Betula*	Cool

Weichselian Glaciation

Glacial deposits of Weichselian age are comparatively little dissected, and
the great end-moraine systems, particularly those of the Brandenburg, Franfurt
and Pomeranian belts, are barely modified. These overlap, and it is clear
that maximum glaciation was not synchronous everywhere. Like the Würm
Glaciation of the Alps numerous closed depressions lie on the surface of the
Weichsel drifts, and, at least in Denmark and Germany, loess does not form a
covering layer. In the major valleys various 'Low Terraces' are related to
this glaciation.

The Stettin moraine (Fig. 2-7) is thought by some to be earlier than the
maximum advance. Its 'overridden' character is generally accepted, but some
dispute obtains on its exact age. Woldstedt (1967) placed it in the Late
Weichselian, prior to its being overridden by the Brandenburg advance. Others,
however, regard it as being older and assign it to the Early Weichselian :
in Poland the Stettin drift is associated with Mousterian artifacts, while in
the European U.S.S.R. similar deposits (the Kalinin Drift) are also deemed
to be Early rather than Late Weichselian.

Whereas initial attention was concentrated on the area glaciated by the
Weichselian Ice Sheet, it was later shown that a more or less complete record
of the entire cold stage existed in the Netherlands, which lay in the
periglacial zone throughout. Supplemented by data from Denmark, the evidence
constitutes a remarkably detailed picture, and is potentially complete in its
major respects.

During the Weichselian the Netherlands became a major centre of loess and
cover sand deposition. Intercalated in these are palaeosols which developed
during the more genial interstadial episodes. Pollen analytical investigations, notably by Zagwijn (1961), Wijmstra (1969) and van der Hammen et al
(1971), demonstrated the contemporary vegetation. During the cold episodes,
however, vegetation appears to have been absent, or at best sparse. Such
periods of 'Polar Desert' conditions are also characterised by ice-wedge
casts demonstrating widespread permafrost. The interstadials recognised are
shown on Figure 2-10. The type localities of the Amersfoort, Moershoofd,
Hengelo and Denekamp interstadials lie in the Netherlands; whereas those of
the Brørup, Bølling and Allerød interstadials are in Denmark (Iversen 1947);
that of the Odderade interstadial is in Germany (Aversdieck 1967).

Weichselian late-glacial time is subdivided into Pollen Zones: (I) Older
Dryas, (II) Allerød, (III) Younger Dryas, based on type localities in
Denmark (Chapter 6). Radiocarbon dating of the zonal boundaries has
established a precise framework for correlating stages in the dissolution of
the Late Weichselian Ice Sheet. For example, Figure 2-11 shows the position
of the ice-margin during Younger Dryas Time (ca. 11,000 to 10,000 yrs BP).
Different sectors of the margin, even though the end-moraines are· continuous
for considerable distances, were not necessarily deposited simultaneously
(Aarseth and Mangerud 1974). The Herdla Moraines of Western Norway are
linked with the Ra Mountains of Eastern Norway, the Cental Swedish End-
Moraines, and finally the Salpausselkä Moraines of Finland.

Elaboration and assessment

Because of its morphostratigraphical character the classical North European
glacial-interglacial classification if inherently defective. Indeed it could

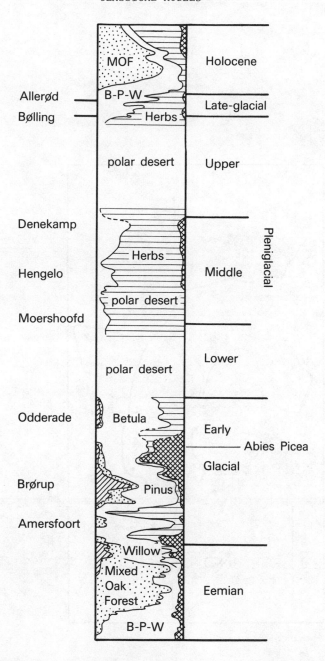

Fig. 2-10. Vegetational history of Europe for the Eemian Interglacial, Last Glaciation and Holocene. Interstadials on left side (after van der Hammen *et al* 1971)

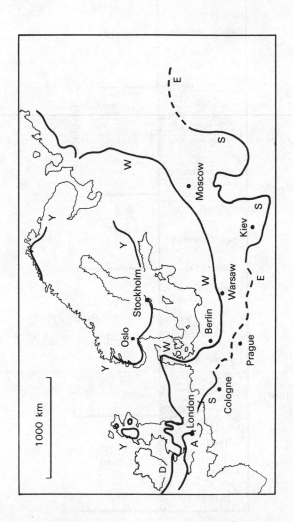

Fig. 2-11.   Maximum extent of Glaciation in Europe. E : Elster. S : Saale.
A : Anglian. Extent of Last Glaciation. W : Weichsel. D : Devensian.
Younger Dryas Ice Margin : Y

only function on the unproven assumption that the oldest moraines lie
farthest south, and northwards successively younger ones occur. The possibility that some end-moraine ridges were subsequently overriden, a matter of
current controversy, highlights some of the difficulties. It is by no means
certain, however, that glacio-tectonics, and 'ground-moraine' on moraine
ridge crests (Duphorn *et al* 1973) demonstrate readvance and overridding.

There can be no doubt that early investigators were hampered by a lack of
data. This is particularly true of knowledge concerning deposits at depth,
for in parts the Quaternary of Northern Germany, for example, may reach up to
500 m in thickness. The initial classification, therefore, necessarily
relied on surface, and near surface exposures, to supplement the morphostratigraphical outline established from the end-moraine systems.

Recently deep bore-hole data has complicated the picture considerably. For
example, in North Germany two 'Eemianp marine transgressions are known to
have occurred : one previous to, the other after the Warthe Glaciation
(Wiegank 1972). A corollary to this is that 'Eemian' interglacial deposits
lying in kettle holes on Saale (Drenthe) end-moraines, may not be the same age
as those similarly located on Warthe end-moraines, despite apparent floristic
similarities (Kukla 1975). From this it would seem that the interval between
Drentne and Warthe Glaciations is of interglacial rank. Kukla (1975) has
indeed argued that such is the case, and in addition has made out a
compelling argument for the recognition of a third 'Eemian' interglacial.

Similarly two Holstein Interglacials are now known to exist (Chapter 6). The
younger, known as the Wacken or Domnitz Interglacial, is clearly separated
from the classical Hoxnian Interglacial by beds which accumulated in a
subarctic climate (the Mehlbeck or Fuhne 'glacial' interval). Pollen
analytical data shows major similarities between the two interglacials :
Azolla filiculoides was common to both, but *Abies* is missing from the Wacken
Interglacial.

Conclusion

A major modification of the classical schemes is required. It is too early,
however, to regard it as pending. Adequate resolution of the confused state
of North European glacial stratigraphy would seem best served by investigation
of long sequences (Chapter 6), by correlation, via terrace systems, with
loess profiles; and by independent dating techniques such as palaeomagnetic
stratigraphy.

THE BRITISH ISLES

The glacial-interglacial sequence starts at the top of the Cromer Forest
Bed 'Series' in Norfolk at a point where Clement Reid (1890) would have
drawn the base of the Pleistocene. Pre-Cromerian Early Pleistocene deposits
in East Anglia ('Crags') do not show any sure signs of glaciation, despite
the identification, by electron microscopy, of quartz grains with surface
textures reminiscent of glacial abrasion.

Undoubtedly the cornerstone of the succession lies in East Anglia, where, in
modern times, the scheme of West (1963) effectively summarised the existing
state of knowledge. It is based primarily on palaeobotanical invesitgation
of interglacial deposits which are used to subdivide the sequence. As such
it was a direct fore-runner to that of the Geological Society of London

TABLE 2-5 Recommended stratigraphical table of the British
           Quaternary. Based on (Mitchell *et al* 1973) with
           additions.

Stage	Type-locality	Boundaries (base)	European correlation
Flandrian		Base of Pollen Zone IV, ca. 10,000 BP	
Devensian	Four Ashes (pit)	Late : 26,000 to 10,000 BP Middle : 50,000 to 26,000 BP Early : pre 50,000 BP	Weichselian
Ipswichian	Bobbitshole	Base beginning of Pollen Zone I I	Eemian
Wolstonian	Wolston (pit)	Base of Baginton-Lillington Gravels	? Saale
Hoxnian	Hoxne (pit)	Base of Pollen Zone H I	Holstein
Anglian	Corton (cliff)	Base of Cromer Till	? Elster ? Saale
Cromerian	West Runton	Base of Pollen Zone C I	?
Beestonian	Beeston (cliff)	Base of Pollen Zone Be I	
Pastonian	Paston (foreshore)	Base of Pollen Zone P I	
Baventian	Easton Bavents (cliff)	Base of Pollen Zone L 4	
Antian		Base of Pollen Zone L 3 LV (forams)	
Thurnian	Ludham (borehole)	Base of Pollen Zone L 2 LIII (forams)	
Ludhamian		Base of Pollen Zone L I LI (forams)	
Waltonian	Walton-on-the-Naze (cliff)	Base of older Red Crag	

(Table 2-5) which attempted a stratigraphic refinement of the previous
nomenclature by naming type-sites (stratotypes) for each stage (Mitchell
et al 1973). More precisely these are boundary stratotypes (Chapter 4) for
the base of each stage is defined. Different methods of subdivision are
used for cold and temperate stages : cold stages are characterised by
lithological divisions, and temperate stages by biostratigraphy (Chapters
4 and 7).

Cromerian Stage

Cromerian Interglacial deposits are known principally from Norfolk where
pollen analytical investigation of the celebrated Cromer Forest Bed 'Series'
has shown it to consist of three separate stages : Pastonian (temperate),
Beestonian (cold), and Cromerian (temperate) (West and Wilson 1966). Farther
south, in Suffolk, a palaeosol, Rubified Sol Lessive, overlies Beestonian
fluvial deposits of an early River Thames (Rose and Allen 1977). Elsewhere
only two other sites are known. At Sugworth, near Oxford, fluvial deposits
of the Thames lie 40 m above the present river. In Somerset the cave
deposits at Westbury-sub-Mendip are believed, on palaeontological grounds
(Chapter 6), to be somewhat younger than the Cromerian type-site in Norfolk.
(Bishop 1974).

Anglian Stage

Like its presumed correlative in continental Europe, the Elster Glaciation
(Table 2-5), the Anglian Glaciation is not delimited by an end-moraine system,
but is known from the distribution of till. Unlike the Elsterian, however,
Anglian outcrops are extensive. At Corton Cliff, near Lowestoft, the
succession at the type site is :

TABLE 2-6   The succession at Corton Cliff, Suffolk
            (Banham 1971)

Bed	Environment	Nomenclature	
7 Plateau Gravels	outwash	)	)
6 Pleasure Gardens Till	flow-till	) Lowestoft Stadial	)
5 Oulton Beds	lacustrine	)	)
4 Lowestoft Till	till	)	) Anglian
3 Corton Sands	estuarine	) Corton Interstadial	)
2 Cromer Till	till	Gunton Stadial	)
1 Silt, clay and mud	estuarine and freshwater	Cromerian Interglacial	

The two principal glacial units, Cromer and Lowestoft Tills, are the product
of different ice-sheets : ice moving on-shore from the North Sea deposited
the Cromer Till, while the Lowestoft Till was derived from the west. Beds 5
to 7 (Table 2-6) are related to deglaciation of the Lowestoft Advance.

Between the Cromer and Lowestoft Tills lie the Corton Sands. Formerly
described as marine, they are now known, on account of their sedimentary
structures, flora and foraminifera, to be estuarine and backwater environment
deposits. Moreover, ice-wedge casts confirm the floral evidence in
demonstrating periglacial conditions (West and Wilson 1968). Thus the
Corton Interstadial (Mitchell *et al* 1973) is so named only on the basis of
a withdrawal of glacier ice and existence of periglacial conditions : there
is no evidence for climatic amelioration. Both Cromer and Lowestoft
Advances are thought to be of Anglian age : because the Cromer Till is
underlain by Cromerian Interglacial deposits, and the Lowestoft Till is
everywhere overlain by Hoxnian Interglacial deposits. This supposition
rests on the assumption that the correct number and order of interglacials
is known; and although perhaps unlikely, the possibility remains that there
could be a considerable hiatus either at the base, or top of the Corton
Sands or at both points.

In the absence of suitable evidence outside East Anglia correlation with
glacial deposits elsewhere is fraught with uncertainty. The Anglian
Glaciation may be represented by the Plateau Drift of the Oxford region,
the 1st Welsh drifts of the Midlands, and by analogy with East Anglia,
those deposits related to the maximum extent of Pleistocene glaciation in
the Celtic Sea as in North Devon and the Isles of Scilly (Fig. 2-13).

Hoxnian Stage

It is widely agreed that the Lowestoft Till at Corton may be traced inland to
Hoxne in Suffolk, the type-site of the Hoxnian Interglacial. There, in a
hollow on the Lowestroft Till (Fig. 2-12) interglacial lacustrine deposits
accumulated Overlying the lake beds are fluvial deposits, probably related
to the terrace system of the adjacent River Waveney (Gladfelter and Singer
1975), while the sequence is capped by periglacial gravels, thought at one
time to contain Gipping Glaciation (Wolstonian) erratics (West 1963).

The vegetational succession of the interglacial was established by West (1956),
though unfortunately, and presumably due to erosion, the declining phase of
the temperate period is missing. It is, however, present at Marks Tey in
Essex (Turner 1970), where the full interglacial cycle is known. There, on
the basis of organic varves, its duration has been estimated as 20,000 to
25,000 years (Turner 1975).

On the basis of a similar vegetational succession, correlation has been
effected with lacustrine deposits in the Birmingham area, similarly
sited on higher ground. But long distance correlation between the Hoxnian
of East Anglia and the Gortian Interglacial of Ireland is unconvincing when
the considerable difference between the second half of each respective cycle
is appreciated (Chapter 6).

A feature of Hoxnian deposits is their occurrence on hill-tops and interfluves
in situations unrelated to present drainage lines. This contrasts with
deposits of the later Ipswichian Interglacial which are found in valleys cut
below the level at which Hoxnian deposits lie. The excavation of these post-
Hoxnian valleys is thought to have taken place at the end of the Wolstonian
Cold Stage, for, in the Midlands, Wolstonian deposits are similarly
preserved on the higher ground only, and do not occur on the valley floors.

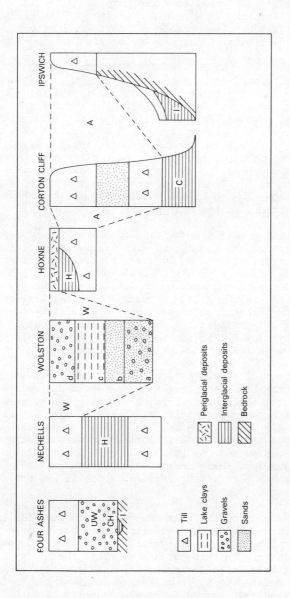

Fig. 2-12.   Diagram of British Stratotypes.   C : Cromerian. A : Anglian. H : Hoxnian.
W : Wolstonian. I : Ipswichian. D. : Devensian. CH : Chelford Interstadial. UW : Upton
Warren Interstadial.   At Wolston - a : Baginton-Lillington Gravels. b : Baginton Sand.
c : Wolston clay. d : Dunsmore gravel.

Wolstonian Cold Stage

Formerly designated the Gipping Stage (West 1963), this has been renamed
Wolstonian, and the type area transferred from the Gipping Valley in
Suffolk, to a type- site at Wolston, Warwickshire, in the English Midlands
(Mitchell *et al* 1973) (Fig. 2-12). The reasons for this are instructive
and illustrate some of the difficulties in classifying glacial deposits.

Till of the Gipping Glaciation was thought to be differentiated from
Lowestoft Till first, by lithology, despite the fact that 'the tills are
nowhere superimposed in a convincing fashion' (West 1963); and second, by
different stone orientations (West and Donner 1956). The identification of
such different till fabrics, however, relied on the supposed lithological
contrasts, so that the argument was partly circular. Solomon (in West 1963)
voiced early objections to it as a means of differentiation, and in addition,
pointed out that a stratigraphic demonstration of the reality of a Gipping
Glaciation was not possible anywhere in East Anglia : no Hoxnian deposits
were known to be overlain by till. These objections were vindicated when,
after extensive field mapping and bore-hole investigation, it was shown that
only the Lowestoft Till existed (Bristow and Cox 1973).

Bristow and Cox (1973) suggested that because the Lowestoft Till represented
the maximum glaciation, it should be correlated with the Saale Glaciation of
Europe, and not, as palaeobotanical data would have it (i.e. post Hoxnian =
Holsteinian), with the Elster Glaciation (Table 2-5). Further, that both
Hoxnian and Ipswichian belonged to the *same* post-Saalian interglacial, being
separated only by a cool phase in the middle of such an interglacial (no
reason was given why 'Hoxnian' should antedate 'Ipswichian' in their scheme).
They also showed that the geomorphological distinction between Hoxnian and
Ipswichian deposits, whereby only Ipswichian beds lie in present valley
systems, was invalid, and cited instances where both so-occurred.

While the burden of their criticism on the 'Gipping Till' has been accepted,
and only one unitary till, the Lowestoft (Anglian), recognised, important
objections have been raised to their other points. Two in particular are
pertinent: first, by palaeobotanical definition, the cool phase at the close
of the Hoxnian, and the similar one at the beginning of the Ipswichian, show
the existence of a cold stage between, even if not represented locally by
glacial deposits (West in Bristow and Cox 1973); second, while Hoxnian
deposits do occur in some valleys, these are *tunnel valleys* excavated
subglacially during the Anglian Glaciation; Hoxnian deposits are not found
in other valley systems (West 1973).

While the fundamental defficiency of the East Anglian region is a failure to
demonstrate a post-Hoxnian/pre-Ipswichian glaciation, this does appear to be
satisfied in the Birmingham area where till overlies interglacial deposits
biostratigraphically correlated with those at Hoxne. Yet, the new type-site
for the Wolstonian Stage at Wolston in the valley of the Warwickshire Avon
(Mitchell *et al* 1973) is unrelated to any such interglacial marker horizon.
It only shows, though admittedly in some detail, the deposits of one
unitary cold stage (Shotton 1953) (Table 2-7).

TABLE 2-7 The type Wolstonian (Shotton 1953)

Unit		Environment
Dunsmore Gravel		outwash
Wolston Clay	Upper Wolston Clay	lacustrine
	Wolstan Sand	sand
	Lower Wolston Clay	lacustrine
Baginton Sand		fluvial
Baginton-Lillington Gravels (with cold fauna)		fluvial

Subsequently, because the succession at the Wolston Pit does not show the sequence of events in full, due to locally variable and diachronous litho-facies (e.g. the Lower Wolston Clay passes laterally into till elsewhere), Shotton (1976) redefined the subdivisions of the stage in terms of what mounts to an areal stratotype (Chapter 4). This incorporates the mapping of Rice (1968), and is applicable to the country between the Wreak Valley and Moreton-in-Marsh. But despite this new extended definition of the Wolstonian subdivisions, it remains one without clear connection to either the preceeding or subsequent interglacial stages : considerable lacunae could occur before and after the stage as presently defined on lithological grounds.

TABLE 2-8 Revised nomenclature for the divisions of the
           Wolstonian (Shotton 1976) (Compare with Table
           2-7)

Member	Equivalent
Dunsmore Gravel	
Upper Oadby Till	Upper Wolston
Lower Oadby Till	Clay
Wolston Sand	Wigston Sand
Bosworth Clays and Silts	Lower Wolston
Thrussington Till	Clay
Baginton Sand	Thurmaston Sand
Baginton-Lillington Gravel	and Gravel

Ipswichian Stage

Bobbitshole, in the Belstead Brook valley near Ipswich, the type site for
this interglacial is, geographically, far removed from deposits of the
preceeding cold stage. A considerable unconformity occurs at its base, and
local glacial deposits, below the level of which the valley is excavated
(Fig. 2-12), are Anglian in age rather than Gipping (Wolstonian) as was
thought when the site was first proposed (West 1963). The beds are overlain
by a periglacial gravel of presumed Devensian age (Fig. 2-12). While Lelief
was affirmed in the post Hoxnian age of these interglacial deposits, the
Geological Society Report showed awareness of the unsuitability of the site by
stating that it might prove necessary to move it to a more appropriate area
(Mitchell *et al* 1973).

The vegetational succession at Bobbitshole is dissimilar to that of the
Hoxnian Interglacial; but it records only the first half of an interglacial
cycle. Indeed the complete vegetational succession for the entire cycle is
unknown from any one site (Fig. 6-5, Chapter 6) : the most complete is from
Swanton Morley in Norfolk (Phillips 1976).

Without a secure type site, as a basis for comparison and correlation,
fragmentary pollen analytical information from elsewhere may be incorrectly
and misleading classified as Ipswichian (see Fig. 6). There may, in fact,
be more interglacials than the resolving power of this particular tool is
capable of showing at the present. From a different point of view it has
been argued that 'The so-called Ipswichian seems to be especially over-
burdened with deposits of diverse ages' (Sutcliffe 1975); more specifically,
'that there are too many distinct mammalian assemblages regarded as Ipswichian
to include in one interglacial' (Sutcliffe 1976). The case for a further
interglacial on this basis, between Hoxnian and Ipswichian has been argued
by Kurten (1968) and Sutcliffe and Kowalski (1976), although this is not
considered likely by Stuart (1976) (Chapter 6).

Devensian Stage

As in North Germany early attention was paid to terrain glaciated during this
cold stage. Much controversy existed on the extent of the Last Glaciation,
and whether it was dual or not, with possible early and late advances. The
discovery of the Four Ashes site near Wolverhampton (Shotton 1967b) showed
that an Early Devensian Glaciation could no longer be sustained : hence only
one, late Devensian, advance is currently recognised. Its limits have been
controversial (Fig. 2-13) but in recent years, based largely on stratigraphic
controls, a measure of agreement has emerged (Institute of Geological
Sciences Quaternary Map of Great Britain 1977).

A great deal has been discovered about the periglacial and interstadial
climates of the Early and Middle Devensian, especially by Coope's investigations
of coleopteran faunas (Coope 1975). This work has enabled mean July
temperatures to be inferred for the Upton Warren Interstadial Complex
(broadly synonomous with Middle Devensian); and has also confirmed the
palaeoecological inferences from pollen data about the Chelford Interstadial,
correlated with the Brørup Interstadial of Denmark.

Whereas until quite recently it was customary to portray several readvances
of the Devensian ice margin, more critical appraisal of the evidence has shown
most of it to be invalid : e.g. Lammermuir line and Perth Readvance are no
longer accepted (Sissons 1974c). The one clear readvance, or more correctly

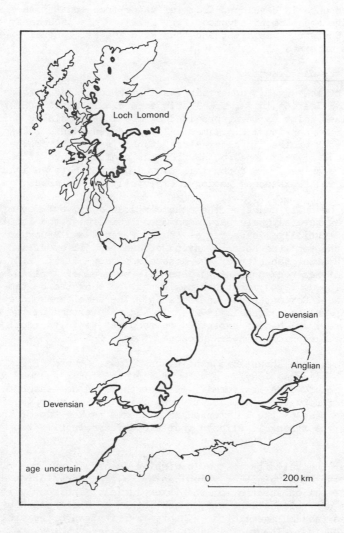

Fig. 2-13.   Principal glacial limits of Britain. Maximum
glaciation in the east is the Anglian, but is of uncertain age
in the west.   The Wolstonian ice-margin is not known

'advance', because it developed from fully ice-free conditions some 12,000
yrs BP, is the Loch Lomond Advance (Fig. 2-13). This is broadly contemporaneous with Younger Dryas time, and thus with the Ra, Central Swedish and
Salpausselka moraines of Scandinavia.

Elaboration and assessment

As is shown by the several uncertainties in the foregoing discussion the
British classification is far from satisfactory. It rests primarily on a
number of fixed point datings, provided by the interglacials, which when
combined, form a continuum of climatically inferred events. Climatic rather
than actual rock cycles, in clear stratigraphic superposition, are the basis
of the scheme. If this were not so then the confusion over the Gipping Till
could not have arisen. No doubt this *status quo* reflects an inadequate
knowledge of the Pleistocene geology, especially in key areas.

Experts will be fully aware of the scheme's drawbacks, but for others,
especially the several Quaternary consumer disciplines, the labels carry
dangers as potentially misleading as those of Penck and Bruckner's. Inevitably
'count from the top' (or bottom) dating procedures followed, and will follow
again, the apparent security of the scheme's pigeon holes: e.g. as far afield
as the Isle of Man rock units in a bore-hole were named Anglian, Hoxnian and
Wolstonian (Thomas 1976) : but how could they be shown to be time equivalent
to the strata at Corton, Hoxne and Wolston? The same terms have been used
around the Irish Sea (Mitchell 1972). Indeed when extending the original
scheme on a regional basis, experts were required to 'fit' local sequences
into the stratigraphic framework proposed (Table 2-5).

When such cycles are shown as a continuous climatic curve (Fig. 2-14) the
effect is one of consolidating a view of earth history, and one which invites
new data to be added on its terms. In fact there is no reason to suppose that
the climatic fluctuations were continuous : it is, for example, impossible to
show that the base of the Wolstonian is the same as the top of the Hoxnian.
There exists the strong likelihood that a considerable time gap separates the
two.

Conflict over the validity of the Ipswichian as a unitary interglacial
exemplifies the uncertainties which can arise from correlation on the basis
of incomplete pollen spectra for the interglacial. Indirectly it points to a
more serious defficiency inherent in the method of correlation by means of
facies-floras (assemblage floras - Chapter 6), instead of by evolutionary
changes in species. The fact is that similar constellations of species were
repeated several times in the Pleistocene, though not perhaps in the same
relative abundance. A regional differentiation of such assemblages could
lead to confusion between interglacials, for in neither the Hoxnian nor
Ipswichian is the regional variation of the flora known throughout Britain.

Conclusion

Sound progress will only follow from an increased knowledge of the actual
Pleistocene geology and its spatial variation. Particularly pressing is the
need to establish the relations between the coastal exposures in East Anglia
with those inland, thence westward, by way of the Breckland, to the English
Midlands. Only when armed with a full statement of the Pleistocene rocks
will meaningful progress be possible : the case of the Gipping Till is a clear
warning of the pitfalls without such knowledge.

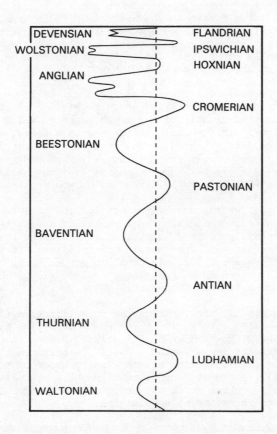

Fig. 2-14.   Schematic climatic curve of the British Quaternary
(after Shotton 1968).   Palaeomagnetic data (Zagwijn 1975) shows
considerable gaps in the earlier Pleistocene (pre Cromerian).
Interglacials on right (Flandrian = Holocene).   Glacials on
left

CENTRAL NORTH AMERICA

The enormous ice sheets which covered Canada and the northernmost parts of the
United States originated in three principal regions : Cordilleran Canada and
Alaska, Keewatin (west of Hudson Bay) and Labrador. In addition, the higher
ground of the Cordilleran system in the Western United States, and parts of
the northern Appalachians, nourished discrete ice caps and glaciers. It is
partly such a contrast in source area, topography, and consequent style of
glaciation, which makes correlation difficult between, for example, the
western United States and the central lowlands.

The Central North American lowlands around the Great Lakes consist largely of
horizontally bedded sedimentary (often carbonate-rich) rocks which contributed
significantly to the predominatly fine-grained deposits which comprise the
extensive till plains of the region. In the mountain areas, with greater
relief, and an availability of greater quantities of quartz-rich debris, the
glacial heritage is somewhat different in style. This contrast is further
heightened by the relatively well preserved state of the sediments of the
lowlands as compared with the highly dissected upland remains.

It is, however, in the Central North American lowland that the classical
scheme of glacials and interglacials was formulated, and in this the names
of two mighty pioneers figure prominently, Chamberlain and Leverett (Table
2-9). The glacial sequence of the western mountain areas is considered in
passing later (Chapters 5 and 10). It could be argued that, because of the
availability of radiometric dates for glaciations previous to the Wisconsin
Stage, the Western area is more important.

Considering that nearly all the terms of the classification (Table 2-9) were
proposed during the last century, it is somewhat surprising that they have
persisted for so long without major modification. Deevey (1965) wrote of
'the coming stratigraphic reassessment', but it is only recently that signs
of it are emerging. Classification was originally based on till sheets (rock
stratigraphic units), landforms (morphostratigraphy), and palaeosols (soil
stratigraphic units). Unlike Northern Europe pollen analysis has not,
hitherto, proved a useful means of subdivision.

Nebraskan Stage

Flint (1971) stressed the need for a revision of this stage for, despite its
rank as a major component of the classification, it is inadequately known.
Originally proposed on the basis of a till in Iowa (Table 2-9), which was
earlier thought to be Kansan, and the stratigraphic relations of that till to
known Kansan deposits, the type locality was subsequently moved to the eastern
part of Nebraska, where till and outwash deposits occur. Clarity of
definition has been hampered because the majority of deposits are buried, a
situation which has led to the unfortunate tendency of labelling 'Nebraskan'
any till below alleged Kansan deposits.

To exemplify the present uncertain state of knowledge : it has been now shown
that Nebraskan till, in its type area above Florence, Nebraska (across the
state line from Iowa), is to be correlated with the upper (Kansan) *rather
than* with the lower (pre-Kansan?) till of the Afton area (Table 2-9). In
turn the lower till of the Afton area is to be correlated with a till in
eastern Nebraska believed to be early Kansan in age. This till overlies a
silt, of late Nebraskan age, which contains volcanic ash dated at 1.2 Ma

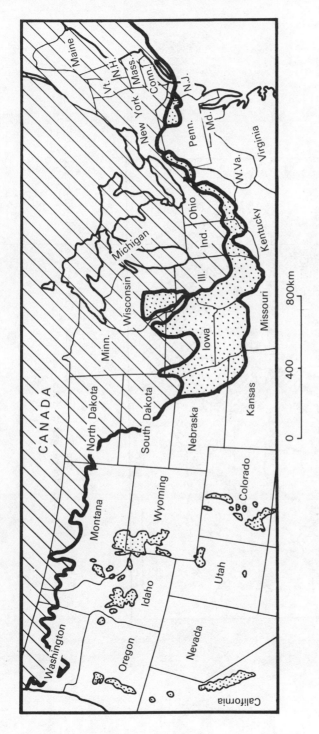

Fig. 2-15. Limits of the Wisconsinan Laurentide and Cordilleran ice sheets (shading), and ice caps (stipple) of the western U.S.A. Pre-Wisconsin glaciated area in Middle West stippled (after Flint 1971)

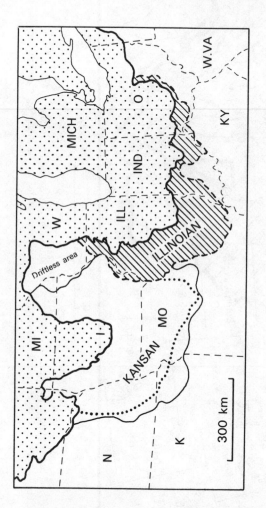

Fig. 2-16.   Extent of the four main glaciations in central North America.
Wisconsin : stipple. Nebraskan : dots (after Flint 1971)

TABLE 2-9   The glacial-interglacial sequence of Central
            North America

Stage	Source[1]	Type-site/area
Wisconsin	Chamberlain 1884, 1895 Leverett 1899	State of Wisconsin : initial description based on end-moraines[2]
Sangamon	Leverett 1898	Sangamon County, Illinois, where Sangamon soil lies between loess (Wisconsin) and till (Illinoian)
Illinoian	Leverett 1896	State of Illinois (Illinois Lobe of Larentide ice sheet) : all deposits between the Yarmouth and Sangamon soils[3]
Yarmouth	Leverett 1898	Yarmouth, Des Moines County, Iowa : soil separating Kansas and Illinoian glacial deposits
Kansan	Chamberlain 1894	Upper till[4] at Afton Junction Pit, Union County, Iowa
Aftonian	Chamberlain 1894	Gravels[5] at Afton Junction Pit, Union County, Iowa
Nebraskan	Shimek 1909	Lower till[6] at Afton Junction Pit, Union County, Iowa

1   Originator of term named.

2   Since defined as commencing at the base of the Roxana silt which rests on
    the A horizon of the Sangamon Soil by Frye *et al* 1968.

3   Redefined in Illinois at the Tindall School section (Willman and Frye 1970).

4   Upper till correlated with surface deposits in the State of Kansas. Kansas
    till redefined in Kansas by Frye and Leonard 1952.

5   Because the gravels were later suspected to represent outwash, transference
    of the term to accretion gley soils on Nebraskan till in southern Iowa
    (Nebraskan 'gumbotill') has taken place.

6   Extends into the State of Nebraska.

(fission track date). Other dates (Boellstorff 1973b) show that at least
five different tills exist in South Dakota and Iowa, spanning the period
between 1.2 Ma and 700,000 years, all of which are called Nebraskan.
Further, that the oldest date for glacial sediments in central North America,
which may approximate the initiation of continental glaciation in that
province, is ca. 2.0 Ma. This comes from Meade County, Kansas, and relates
to the dating of glacial sands and gravels (Boellstorff 1973b).

The increasingly demonstrated complexity of the Nebraskan stage endorses
Dort's (1972) suggestion that glaciation was multiple, but whereas he
advocated only two separate advances, the actual situation may be more
intricate.

Aftonian Stage

Sands and gravels of the original type Aftonian (Table 2-9) were later shown
to be fluvioglacial outwash deposits. The stage is now based on a well
developed palaeosol, the Aftonian Soil. This frequently occurs, with a well
developed B horizon, as a deep soil profile on Nebraskan till : for example,
at Iowa Point, Doniphan County, Kansas. Other Yarmouth soils include accretion gleys which, according to previous nomenclature and concepts, were called
'gumbotills' : that is, soils which had developed *in situ* on till. A limited
amount of faunal data allows inference of semi-arid climates in south-west
Kansas.

Kansan Stage

This was defined from the upper till of the Afton Junction area, Iowa (Table
2-9), which could be shown to be the same age as the surface drift of north-
east Kansas. The end of the stage is marked by the Yarmouth Soil. Stratigraphic-geomorphic relationships show that major valleys such as the
Mississippi, Ohio and Iowa were already entrenched by this time; but further
erosion, post-dating the Yarmouth Soil, effected considerable dissection of
the Kansan tills.

In Illinois a subdivision of the Kansan is possible, and is shown on the
accompanying time-distance diagram (fig. 2-17). Ice moved into Illinois from
both western and eastern sources. That from the east was two-pronged : one
lobe advanced from the Lake Erie area, the other from the Lake Michigan area
(Johnson 1976). Although these deposits of eastern provenance may be clearly
distinguished from those of the west, chiefly on mineralogical grounds, age
correlation between the two has not yet been positively demonstrated, it is
merely assumed. Of the two, the western drift is most readily correlated
with the type Kansan, for it is coextensive with till of that stage in north-
east Kansas and adjacent Iowa. Somewhat problematical is the Hegeler Till
(fig. 2-17), mineralogically distinct from younger ones, but of unknown age :
it could be pre-Kansan.

Phases of advance and retreat are inferred from outwash deposits, but also
from fossiliferous silts. These have yielded molluscs indicating a northern
temperate climate (Leonard *et al* 1971), but whether this was interglacial as
opposed to interstadial is unresolved. Lack of evidence for contemporaneous
weathering has been taken to indicate a brief period of time, so that the
deposits are unlikely to be interglacial (Johnson 1976) believed that
several advances had taken place, but that it was not possible to elucidate
the nature of the intervals between them. There is no doubt, however, that

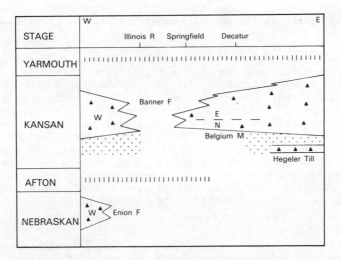

Fig. 2-17. Time-distance diagram for
the Nebraskan and Kansan in Illinois
(after Willman and Frye 1970 and Johnson
1976)

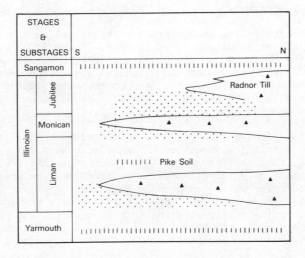

Fig. 2-18.   Time-distance diagram for
the Illinoian in Illinois (after Willman
and Frye 1970 and Johnson 1976). Till :
triangles. Loess : stipple. Soils :
vertical lines

several Kansan tills exist, and work in western Illinois and in Kansas has
shown that the record is rather more complicated than assumed hitherto
(e.g. Boellstorff 1973a). In Indiana multi-till sequences have been
recorded, though their dating can only be established as pre-Illinoian
(Teller 1972).

Two unrelated and independent means of dating allow the Kansan Stage to
be placed in approximate position in the Pleistocene time-scale : (1) in
eastern Illinois a proglacial lake silt deposit (Belgium Member in fig. 2-17)
is magnetically reversed, hence is older than 700,000 years, (2) a layer of
tephra (volcanic ash), the *Pearlette Ash*, interbedded in the Sappa Formation
of Kansas and Nebraska, which consists of alluvium and loess of late Kansan
or earliest Yarmouthian age, allows dating by tephrochronology (Chapter 5).
The Pearlette Ash was formerly thought to represent merely one event, but it
is now known that three such erruptions occurred, the one represented in the
Sappa Formation being the Pearlette Ash O type, which is about 600,000 years
old (Izett *et al* 1972). These two datings have a considerable bearing on
the classification of the glacial deposits of North America.

Yarmouth Stage

Leverett (1898) defined the Yarmouth soil from a well section near Yarmouth,
Des Moines County, Iowa, as representing the interval of weathering and organic
accumulation between the deposition of Kansan and Illinois glacial deposits.
The original section is no longer available, but in nearby Lee County, Iowa,
the Yarmouth Soil, consisting of 4 m of accretion gley, has been described
overlying Kansan till, and being overlain in turn by Illinoian till. The
Yarmouth Soil is usually twice as thick as that of the Sangamon interglacial,
hence is take to indicate that the Yarmouth interglacial was longer, by
possibly three or four times.

Peat deposits, on which pollen analytical investigations have been carried out,
have allowed some climatic inferences to be made. There is doubt, however,
as to the age of the peats, which are claimed to be Yarmouthian because they
lie between tills of alleged Kansan and Illinoian age. Incorrect dating may
thus lead to erroneous conclusions about the contemporary climate of Yarmouthian time. For example, in Wisconsin, former lake deposits originally
classified as Yarmouthian, have now been shown, first by stratigraphic
correlation, and second by radiocarbon dating (26,060 BP) to be Wisconsinan
in age (Black 1976). Indeed Deevey (1965) commented that no plant bearing
beds could safely be called Yarmouthian. Furthermore he was sceptical of
the existence of a Yarmouth Interglacial, and pointed to difficulties in
sustaining the concept of a Yarmouth Soil away from its type area (Deevey
1965). Any such correlation is hampered by the numerous phases of incision
and aggradation which have occurred subsequently, as well as by the facies
variations imposed by the different climatic regions which the palaeosol
must necessarily be subject to. Flint (1971), however, disagreed with
Deevey, and believed the Yarmouth Soil to be at least as extensive as that of
the Aftonian.

Illinoian Stage

According to Leverett (Table 2-9) the Illinoian Stage includes all the
deposits between the Yarmouth and Sangamon Stages, the type area being the
Illinois Lobe of the Laurentide ice sheet. No type section was proposed.
Ice covered almost 90% of the state of Illinois, and deposited till from

several distinctly recognisable sub-lobes. Although a degree of controversy
obtained in the past over the classification of the glacial deposits three
Substages are now recognised (Willman and Frye 1970. Johnson 1976). It has
been suggested that tills of the Liman Substage reflect the incorporation of
considerable quantities of proglacial loess. By and large the till sheet
thins appreciably towards its margin and does not bear a constructional
topography. In north-east Illinois tills associated with this advance may in
fact be older, and of Kansan age (Black 1976), but this is uncertain.
Separating the glacial deposits of this advance from those of the subsequent
Monican Substage is the Pike Soil, a moderately developed palaeosol with a
distinct B horizon. It clearly demonstrates deglaciation and a prolonged
period of weathering : but is it of interglacial or interstadial rank?

Texturally dissimilar to tills of the Liman Substage, those of the Monican
Substage show that the advance was almost as extensive as the earlier one.
Deglaciation was accompanied by the development of extensive stagnant ice,
indicated by the crevasse controlled patterns of fluvioglacial deposits on the
till plain. The soil separating this from that of the Jubileean Substage
advance is immature and has not been named. Contemporary molluscan faunas
show a cool to sub-arctic climate (Willman and Frye 1970) so that the period
may be inferred to have been briefer than the Pike Soil interval.

Though less extensive than those of the earlier advances, the Jubileean
Substage is different because it is characterised by more constructional
topography, with locally prominent moraine-ridges at the margin of the Radnor
Till. Not unnaturally such constructional topography has led to notions that
the advance is considerably younger in age, but the presence of the Sangamon
Soil on its surface discounts such a possibility and shows the Radnor Till to
be Illinoian. Like the Yarmouth Soil, however, this transgresses climatic
boundaries, is subject to considerable variation in facies, hence is open to
misidentification.

Sangamon Stage

Designated by Leverett this stage is represented by a soil which, in north-
west Sangamon Count, Illinois, overlies Illinoian till and is covered by
Wisconsin loess. It extends from its type area, west to the Great Plains,
south to the High Plains of Texas, and east to Ohio. In doing so it crosses
several distinct climatic provinces and changes facies acccordingly : thus in
Indiana, Illinois, Iowa and east Nebraska it is a priarie soil and accretion
gley, whereas as the climate becomes progressively more arid to the west it
changes through cherozem to chestnut, and eventually to a desert soil. In its
type area it is only buried by loess of Wisconsin age, but has been found
buried by till for distances up to 60 km within the Wisconsin ice limit. Its
profile characteristics include a well developed B horizon, up to 1 m thick;
while on calcareous till decalcification to depths of 2 m occur.

Its reputed value as a stratigraphic marker has already been touched upon,
although it is important to note that nowhere has it been dated by independent
means : e.g. by radiometric determination. During the preceding Illinoian
Stage the majority of valleys had been infilled so that in topographic terms
the Sangamon Soil now cuts across them. On stable till plains pedogenesis was
uninterrupted, and on gentle slopes wash-processes contributed to the
accumulation of accretion gley soil profiles (formerly interpreted as
gumbotill).

Other sediments of Sangamon age include organic remains, notably in south-east Indiana where Kapp and Gooding (1964) were able to show a pollen sequence, representing the entire interglacial, between Illinoian and Wisconsin tills. Perhaps the best documented locality, however, is that of the Don Valley Brickyard near Toronto, where pollen analytical investigation allowed inference of somewhat warmer climatic conditions than obtain in that region today (Terasmae 1960).

Wisconsin Stage

Chamberlain (1895) first used the term Wisconsin as a formation, and which he substituted for his earlier version of 'East Wisconsin Formation'. But it was Leverett (1899) who classified the deposits in question as belonging to the Wisconsin Stage. This was based exclusively on the, barely modified, constructional topography of numerous end-moraines which distinguished the glacial deposits so readily from those of earlier glaciations; hence the original designation is a morphostratigraphical one. Since then the term Wisconsin has been extended to include earlier deposits of glacial and non-glacial origin.

The advent of radiocarbon dating, and the thousands of dates produced since 1949, has allowed a refinement of detail unequalled anywhere else in the world, at least for the later part of the stage. These have confirmed earlier concepts of an Early Wisconsin Glaciation and a later 'Classical Wisconsin' Glaciation, the latter being based on the freshness of its topographic landforms, though the term 'classical' is no longer used. Despite the precision of radiocarbon dating the period beyond the range of that remains inadequately known. Recently, based on external criteria the base of the stage was extended from 70,000 years to 100,000 years + (e.g. Dreimanis 1976), which is a measure of this uncertainty. Such a long, as opposed to a short, chronology, does seem more appropriate for external correlations (Chapter 10).

Two areas are exceptionally well known : the state of Illinois, and the area formerly occupied by the Huron, Erie and Ontario lobes of the Wisconsinan Laurentide Ice Sheet.

Illinois. Willman and Frye (1970) presented a systematic outline of the Wisconsin stage in Illinois, but in the short time since then considerable elaboration and further progress has been achieved (e.g. Johnson 1976). The stage is subdivided into five sub-stages : Altonian, Farmdalian, Woodfordian, Two-Creekan and Valderan, a scheme generally adopted for the western areas of central North America.

The base of the Altonian is the contact between the Sangamon Soil and the overlying Roxana Silt. The silt is subdivided into three members (fig. 2-19) which contains three palaeosols : the Chapin Soil, thought on account of its moderately developed nature to represent the longest interval, Pleasant Grove and Farmdalian Soils. Radiocarbon dating has shown that the Meadow Loess Member accumulated between 35,000 and 37,000 BP. There is no radiometric control for earlier events.

Tills of Altonian age outcrop in north-west Illinois, and are also known in subsurface borings. The Capron and Argyle Till Members (pre 41,000 BP) are separated by the Plano Silt Member which accumulated between 41 and 32,000 BP. Capron Till may be dated very precisely for it has been sandwiched by radiocarbon dates of 32,000 and 27,000 BP. Older tills discovered in the

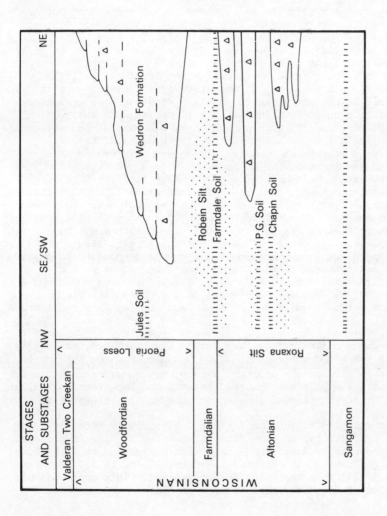

Fig. 2-19. The Wisconsinan Stage in Illinois (after Willman and Frye 1970 and Johnson 1976). Compare with Fig. 2-20. (Ontario and Erie Lobes)

subsurface only, and classified as belonging to the earlier part of the
Winnebago Formation, are of uncertain age.

The Farmdalian substage is based largely on the Robein Silt, and lasted from
28 to 22,000 BP. Major deglaciation of the Altonian ice sheet occurred and
the Farmdale soil developed on the surfaces of both Robein and Roxana Silt.
Gruger's (1972) pollen analytical investigations demonstrated the existence
of boreal forest in central and northern Illinois at this time.

A sequence of tills, outwash deposits and loesses comprise the Woodfordian
Substage. Stratigraphically it extends from the top of the Robein Silt to
the base of the Two Creeks deposits, and covers the period from 22 to 12,500
BP. Moving across an accidented relief the Woodfordian Ice Sheet was differentiated into several distinct lobes the tills of each being capable of
ready mineralogical differentiation. Classification and interpretation by
Willman and Frye (1970) was based on end-moraine ridge (Fig. 8-7), with
some 32 such features being identified in the Lake Michigan Lobe area alone,
and which were built in only 6,000 years. Their regular size and spacing
tends to suggest some sort of climatic periodicity, an explanation favoured
by Willman and Frye (1970). But Wright (1976a. 1976b) showed that local
pollen data, notably at Wolf Creek (Chapter 7), showed unidirectional
climatic change, and was incapable of supporting Willman and Frye's explanation. Instead he favoured repeated surging of the Lake Superior Lobe, and
suggested that the existence of tunnel valleys, indicative of basal meltwater
flow, supported such an hypothesis. Other alternative explanations include
the suggestion that many of the moraines represent a draping of till over
pre-existing topographic eminences (Moran 1971); or that they may be
composed of superglacial till deposited on stagnant ice (e.g. Bleuer 1974).
According to the superglacial till explanation there is no need to postulate
so many readvances and retreats of the ice margin. The recent wave of
investigations on till lithology, (see Johnson 1976 for review) has shown
that older correlation, based on end-moraines, is incorrect (see discussion
Chapter 8). This has yet to acquire independent verification from radiocarbon
dates.

A feature of the younger tills is that they become finer grained, especially
after the period of ice-withdrawal about 17,500 BP, a phenomenon attributed
to the incorporation of lacustrine sediment within readvance ice sheets. A
significant phase of ice withdrawal occurred between the Batestown and Snider
tills (Fig. 8-8) for the surface of the former is oxidised to depths of 1 m.

Extra glacial loess deposits of Woodfordian age, collectively referred to as
Peoria Loess, have been subdivided into four zones based on clay mineralogy.
The base has been shown to be diachronous in Illinois, just as it is in Iowa
(Ruhe 1976). A significant break occurs between 16,500 and 15,500 BP,
between Zones II and III, when the Jules Soil developed. This corresponds
in part to the period of ice withdrawal between the Batestown and Snider
Tills.

The late-glacial includes the Two Creekan and Valderan Substages which are
based on the Two Creeks section on the north-west shore of Lake Michigan
(Table 2-10).

TABLE 2-10   The succession at Two Creeks, Wisconsin

Till	Interpretation	Classification
5 till	Valders Till	VALDERAN
4 stratified fine-grained deposits with molluscs and plant remains	Glacial Lake Chicago. Lake Level raised by readvance	
3 thin peaty layer (forest litter)	Lake level fall	TWO CREEKAN
2 varved silt clay	Glacial Lake Chicago	
1 till	Port Huron Till	WOODFORDIAN

During the retreat from the Late Wisconsin maximum advance, which was inter-
rupted by the minor oscillation when the Port Huron end-moraines were formed,
a major deglaciation was deemed to have occurred during the Two Creeks period
(Table 2-10), dated to ca. 11,850 BP. This was followed by major readvance,
when the Valders ice moved 320 km south as far as Milwaukee. It was so-named
after the red till at Valders, Wisconsin, and at that time it was thought that
all red tills in the region were the same age.

Not unnaturally parallels were sought farther afield for these events : the
Two Creekan was correlated with the Allerød of Europe, despite it having been
impossible to demonstrate a similar event elsewhere in Central North America.
Indeed radiocarbon dating showed that the Two Creeks event actually antedated
the European Allerød.

Eventually extensive stratigraphic and geomorphic investigations required a
revision of the original scheme. It was shown that three red tills occur in
the region, two of Woodfordian age, including the original type Valders Till,
and one of post Two Creekan age (Evenson *et al* 1976). This new data is
summarised in Figure 8- 9 . (Chapter 8)

The Huron, Erie and Ontario Lobes. The Wisconsin Stage in these areas is
exceptionally well known as a result of more or less continuous investigation
since the end of the second world war. This very quickly overturned earlier
views, based on simple till counting, that had been heavily influenced by
European classifications and the four-fold sequence of North America : at one
time, for example, it was thought that all four glaciations were represented.
The present picture has emerged through detailed field mapping (see e.g.
Dreimanis and Goldthwait 1973), lithological, textural, mineralogical, fabric,
trace element, statistical, and other studies, though above all by a radiocarbon dating framework unequalled elsewhere.

Based on general synchroneity of advances in the major ice lobes, a three-fold

subdivision is possible : Early Wisconsin (100,000 ± 53,000 BP), Middle
Wisconsin (53,000 to 23,000 BP) and Late Wisconsin (23,000 to 10,000 BP)
(Dreimanis 1976). This is capable of more detailed subdivision based on
multiple criteria, including water laid deposits which often contain organic
remains. Early and Late Wisconsin were characterised by extensive ice
advances, whereas the Middle Wisconsin consisted of a long cool composite
interstadial. Only the Middle and Late Wisconsin have been reliably calibrated
by radiocarbon dating.

Early Wisconsin time witnessed the deposition of the Becancour Till (Nicolet
Stadial) by ice, in the St. Lawrence lowland, which blocked the drainage so
as to form Lake Scarborough in the Ontario Basin. At Toronto the Scarborough
Formation consists of deltaic sediments containing plants and animals
representative of a cool climate; though beetle remains suggest a somewhat
subarctic climate with temperatures 6° C lower than today (Morgan 1975).
Farther south-west the climate was similar to today's, so that some palaeosols,
formerly classified as Sangamon, may well be Early Wisconsin in age.

Pollen data show that a cool boreal climate obtained during the St. Pierre
Interstadial Radiocarbon dates on the St. Pierre peat, and other related
deposits, are either finite, falling in the range 63,000 to 67,000 BP or
infinite (Chapter 5). In the Atlantic provinces, for example, all the dates
are infinite. Thus, because the finite dates are at the limit of the radiocarbon dating method, it would seem prudent to regard them as minimum
indications of age only : the St. Pierre Interstadial may be considerably
older. Frenzel (1973) suggested as much when he correlated it with the
Brørup Interstadial of Europe (ca. 100,000 BP). The greatest advance of the
Early Wisconsin occurred during the Guildwood Stadial when all the Great Lakes
and St. Lawrence region was ice-covered.

During the Middle Wisconsin the Thorncliff Formation, of Scarborough Bluffs,
Toronto, was deposited in a lake dammed by ice occupying the St. Lawrence
valley. This constitutes the type site for both Port Talbot and Plum Point
Interstadials. Two tills, the Seminary and Meadowcliff, are interbedded in
the upper part of the formation, and demonstrate the proximity of the ice-
sheet, which on occasion reached the Toronto area. The precise age of the
Port Talbot I Interstadial is not know, but a cluster of radiocarbon dates
between 42,000 and 48,000 BP suggest reliable dating for the Port Talbot II
Interstadial (Dreimanis 1976). Pollen data shows that the climate of both
was cool and boreal, though the considerable degree of soil development during
Port Talbot I suggests that it was warmer than Port Talbot II. Well beyond
the ice margin, in Ohio, the Sidney Soil represents all of Middle Wisconsin
time. Advances of the ice sheet during this period were considerably less
extensive than those of the Early or Late Wisconsin, with the strongest one
taking place between the Port Talbot and Plum Point Interstadials (Cherry
Tree Stadial).

Maximum expansion of the Wisconsinan Laurentide Ice Sheet occurred during the
Late Wisconsin in Ohio and Indiana, but not all the lobes reached their
maximum extent simultaneously. Much depended on the subglacial relief and the
dynamics of each individual lobe. A feature of this time was the development
of the direct precursors to today's Great Lakes. Initially ponded up against
higher ground ahead of the advancing ice, retreat allowed them to develop and
widen. Evidence for the lakes is both extensive and impressive, consisting
in the main of cliffs, beach ridges, and a variety of deposits.

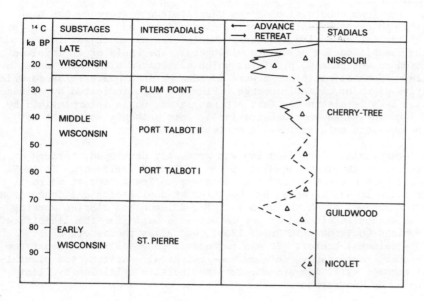

¹⁴C ka BP	SUBSTAGES	INTERSTADIALS	← ADVANCE → RETREAT	STADIALS
20 –	LATE WISCONSIN			NISSOURI
30 –		PLUM POINT		
40 –	MIDDLE WISCONSIN			CHERRY-TREE
50 –		PORT TALBOT II		
60 –		PORT TALBOT I		
70 –				GUILDWOOD
80 –	EARLY WISCONSIN	ST. PIERRE		
90 –				NICOLET

Fig. 2-20   The Wisconsinan in Ontario (after Dreimanis)

Present status of the Central North American Sequence

There are signs that the 'stratigraphic re-assessment' of Deevey (1965) has begun, for clearly each of the different classical stages contains major interpretive problems. This arises due to increasingly sophisticated work being incompatible with the over-rigid classical model which, in retrospect, must have been greatly influenced by thinking in Europe at about the same time as its inception.

Semantic confusion arising over the indiscriminate mixing of geologic-climate terms with lithostratigraphic and chronostratigraphic ones is considered in Chapter 4. Multi-till sequences within the glacial stages as currently defined may eventually require discrimination as being representative of additional stages. Equally palaeosols, now regarded as interstadial, may be reinterpreted as interglacial. The importance of the Sangamon concept seems altogether disproportionate with regard to evidence for it. Nowhere is it dated radiometrically and the practice of labelling the strongest developed of the younger palaeosols in a region 'Sangamon' is singularly ill-advised.

Within, and just beyond, the conventionally accepted range of carbon-14 dating, some of the early dates must be regarded as minima only (Chapter 5). Much of the status of the Early Wisconsin is not so much based on intrinsic evidence, but on comparison with data from elewhere - hence the 'stretching' of the chronology in recent years to accommodate dating of the Last Inter-glacial at 125 ka. Till stratigraphy in the Late Wisconsin is a special problem, and the different interpretations now current are discussed in Chapter 8.

EAST AFRICA

Although the classical subdivision of the East African Quaternary has been
thoroughly, though comparatively recently, discredited, and relegated to the
role of an historical curiosity, it bears examination as an object lesson with
regard to the classification of events elsewhere. The basis of it involved
the recognition of wet phases, pluvials, which alternated with drier ones, or
interpluvials. These climatic terms were founded on inadequate evidence which
lacked positive signs of climatic change. Pluvials were indicated by evidence
of former high lake levels in the East African lakes, while interpluvials by
evidence of lake shrinkage and 'dessication'. Some authorities sought to
correlate these events on a world-wide basis.

The Pluvial-Interpluvial classification was gradually developed, through an
additive process, in the years leading up to the Second World War. In 1947
it was ratified by the first Pan African Congress on Prehistory at Nairobi,
while a year later it was accepted by the International Geological Congress at
London. Thus a scheme, of which, even in 1939, Solomon and O'Brien working
in Uganda, had wished to 'discard it completely as a basis for the classification of the African Quaternary'(O'Brien 1939), was given the sanction of
important international bodies. It was to bedevil progress for at least the
following decade. Regrettably, in some non-geological quarters, the pluvial-
interpluvial concept still lingers on, despite decisive criticism by Flint
(1959), Bishop (1962) and others.

Table 2-11 outlines the principal climatic phases recognised. In essentials it
is not a scheme based on the stratigraphic succession of Quaternary deposits
that have been described and carefully correlated. Instead it represents a
series of inferred climatic cycles which, when presented as a chart gives an
impression comparable to standard geological correlation and stratigraphic
tables. Indeed at various times the units identified have been referred to as
'stratigraphic units' (Leakey 1952), or 'climatic stratigraphic units' (Clark
1957). Clark (1965) went further and used the term 'Gamblian' in three
different, but quite specialist senses: (i) Gamblian deposits, (ii) Gamblian
Pluvial, (iii) Gamblian time. In fact the terms of the classification lack a
stratigraphic definition, they refer to notional climatic cycles, erected at
geographically widely spaced localities, hence are insufficiently objective to
be used as a basis for inference about earth history.

The Gamblian Pluvial was based on strata in Gambles Cave, near Elmenteita,
Kenya (now known on archaeological grounds to be Holocene), and by high shoreline some 155 m above Lake Nakuru. At Olduvai Gorge, Tanzania, the pluvial
is indicated by a wide 'mature' valley, from which it is inferred that greater
discharges than those which obtain today formerly occurred, hence the climate
was wetter. To infer a former climate from a geomorphological features is
unacceptable. The Kamasian Pluvial was based on tuffaceous sand and silt,
gravels and diatomite on the floor of the Eastern Rift Valley. These beds are
greatly deformed by faulting and tilting - amounting in one instance to ca.
600 m. The Kageran Pluvial was based on a river terrace, 82 m above the
Kagera River, which was thought to have been deposited during a period of
increased discharge, hence a pluvial (Wayland 1934). No description of this
site was published, yet it formed one of the bases of an impressive table of
climatic change in East Africa.

Interpluvials were recognised on a similarly insecure basis. The First
Interpluvial was based on 'bone beds' in the Lake Albert Basin, which indicated

Table 2-11 East African Pluvial-Interpluvial Sequence

PLUVIAL – Interpluvial	EVIDENCE	TYPE-LOCALITY
NAKURIAN PLUVIAL	Lake deposits and shorelines	Lake Nakuru, Kenya
dry phase	Red blown sand, Gambles Cave	Gambles Cave, Kenya
MAKALIAN PLUVIAL	Lake deposits and shorelines	Makalia River Valley, Kenya
1st Post-pluvial dry phase	Red soil and blown sand	Gambles Cave and Makalia Valley
GAMBLIAN PLUVIAL	Lake deposits and shorelines	Gambles Cave and Gambles Drift, Kenya
3rd Interpluvial	Palaeosol (red) and lag (alluvial) gravel	Nsongezi area, Uganda
	Unconformity top of Bed 4 Olduvai	Olduvai Gorge, Tanzania
KANJERAN PLUVIAL	Lake deposits and shorelines (some vetebrate remains)	Kanjera area, Kenya
2nd Interpluvial	Red bed (fluvial) Olduvai Bed 3	Olduvai Gorge, Tanzania
KAMASIAN PLUVIAL	Lake deposits (unfossiliferous)	Kamasian Plateau, Kenya
1st Interpluvial	'Bone beds' indicating 'dessication'	Lake Albert Basin Uganda
KAGERN PLUVIAL	Fluvial gravels (unfossiliferous)	Kagera River Valley, Uganda

dessication after the lake level had fallen (Wayland 1934). Such an event,
however, may have been caused by tectonic movements in the rift valley, and
not by climatic change. Red beds were also used as evidence of dessication;
but the interpretation of such deposits has altered radically since the days
when they were used in such a way in East Africa. Now they are readily
attributed to erosion and transportation of red soils in wet climates, prior
to deposition and protection from subsequent reducing processes. Rapid
burial in lake sediments would ensure such protection. Unconformity at the
top of Bed 4 is used to indicate the Third Interpluvial at Olduvai Gorge; but
once again this could have been due to tectonic uplift rather than climatic
change.

Assessment

Thus the Pluvial-Interpluvial sequence, based on climatic inferences from
variable sets of evidence, mostly geographically separate, cannot withstand
detailed scrutiny. In particular it would seem that many indicators of
fluctuating lake levels, and hence by inference climatic change, may be due
to tectonic causes. In any event a classification based on inferred climatic
cycles is quite untenable as a means for subdividing earth history (Chapter 4).

Since 1958 or so, work which combines stratigraphic field mapping, mammalian
palaeontology K/Ar dating of volcanic beds, and palaeomagnetic stratigraphy
has elucidated local sequences greatly (Bishop and Miller 1972). But overall
progress is handicapped by an absence of long continuous sequences of deposits.
The longest available, at Olduvai Gorge, is greatly complicated by a variable
tectonic history which has obscured and rendered interpretation of environmental change difficult (Hay 1972). The same is true of the Omo-Rudolf Basin,
where the record of recent lake fluctuations shows how difficult it is to infer
important climatic changes from such evidence : in 1896 Lake Rudolf was 15 m
higher than it is today, in 1955 it was 5 m lower, while over the past 2,000
years it has fluctuated through a range of 40 m (Butzer 1971).

Unquestionable evidence for repeated glaciation of Mount Kilimanjaro shows
that major climatic changes did occur. Unfortunately, however, detailed
correlation between the highlands and the rift valley sedimentary sequences,
mostly tectonically controlled, does not seem imminent.

Chapter 3
THE OCEANIC RECORD

Whereas the stratigraphical record of Quaternary events is imperfect on the continents, there are places on the deep ocean floors where it is arguably complete for appreciable parts of, and in some instances all of Quaternary time.

That proposition carries with it the assumption that sedimentation would have been more or less continuous. Such a concept is thought to be inherently unlikely by many geologists, who would argue that gaps in the stratigraphical record lasted longer than the actual time spans represented by rocks. The pelagic environment, however, defined as the open ocean, represents a special case. It is exceptional in that given unchanging marine conditions for long periods then sedimentation, by the settling of material through ocean water, may have continued uninteruptedly.

Examination of many deep-sea cores shows that detailed inter-core correlation is possible in terms of their isotopic, palaeontological and magnetic characteristics (e.g. Emiliani and Shackleton 1974). This detailed correlation suggests that the events recorded are likely to be synchronous throughout the major ocean areas, and also that the mode of sedimentation was essentially continuous. Here then is a record potentially valuable for establishing global stratigraphical stages for the entire Quaternary.

SEDIMENTS OF THE PELAGIC ENVIRONMENT

Successful investigation of the Quaternary record depends on the recovery of continuous cores of deep-sea sediment. An earlier generation of corers was able to recover only limited samples. But the advent of the Kullenberg piston corer in 1947 opened up new possibilities. Columns of sediment between 10 and 30 m in length were collected, moreover with high recovery rates. A newer generation of corers promises even better results.

Not all pelagic environments are equally suitable for sampling. Quaternary sediments are absent over extensive areas of the ocean floor, and what remains has often been eroded and redistributed by bottom currents in the open ocean, or by turbidity currents within reach of the continental slope. Ocean basins, seemingly appropriate as sediment traps, are inherently unsuitable because sediment eroded from elsewhere may be redeposited there. Emiliani (1963) suggested that the best locations for undisturbed sediments are the gentle flanks of low submarine mounds. 'Undisturbed' he defined as accumulations of sediment *in situ* which had built up through the settling of material through the open ocean.

Sediment columns with slow accumulation rates tend to have their details obscured by the mixing effects of burrowing organisms (bioturbation). In exceptional cases this can extend through 20 to 30 cm of sediment possibly representing several thousands of years. 4 to 5 cm, however, appears to be

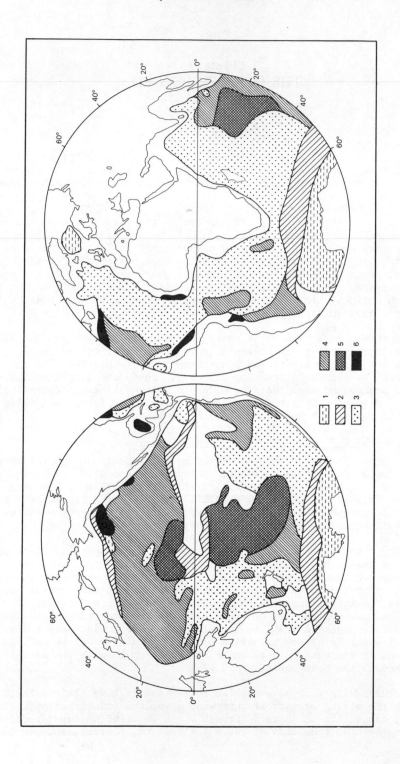

Fig. 3-1. Distribution of deep sea sediments (after Shephard 1973). 1 : Glacio marine. 2 : Siliceous Ooze. 3 : Calcareous Ooze. 4 : Brown clay. 5 : Authigenic. 6 : Turbidites

the normal extent of such mixing (Arrhenius 1963). Rates of accumulation vary
from less than a cm/1000 years to several cms/1000 years (Emiliani 1963).
Sedimentation rates of 5 cm/1000 years or greater are best suited to oxygen
isotope analysis, and such thicknesses mitigate against the blurring effect
of bioturbation (Shackleton 1976). Unfortunately most deep-sea cores which
extend back into the Middle Pleistocene have low sedimentations rates, hence
are rendered less valuable through the effects of bioturbation. Hitherto the
best and most detailed records have come from areas of oceanic upwelling, off
North America and North Africa, but also from the subantarctic of the Indian
Ocean (Shackleton 1977a).

The distribution of the principle pelagic lithologies (Fig. 3-1) is described
and illustrated by Shephard (1973), whose classification, one of many, is
reproduced in its essentials as Table 3-1.

TABLE 3-1 Classification of ocean sediments (based on
           Shephard 1973)

Pelagic deposits

Brown clay (with less than 30% biogenic material)

Authigenic deposits (minerals crystallized in sea
     water, e.g. manganese nodules)

Pyroclasts (tephra)

Biogenic deposits (more than 30% biogenic)

Foraminiferal (*Globigerina*) Ooze, consisting of
     shallow and deep water foraminifera and other
     calcitic material

Chalk (nannoplankton) Ooze

Radiolarian Ooze

Coral reef debris, coral sand and coral mud

Terrigenous deposits

Muds (more than 30% silt and sand of terrigenous origin)

Turbidites and slide deposits

Glacio-marine deposits (iceberg rafting processes)

Oxygen isotope analysis is carried out on the biogenic components of pelagic
sediments, which by and large constitute the various oozes. These contain more
than 30% of skeletal material as well as high percentages of clay. Calcium
carbonate, silicon dioxide and calcium phosphate is taken out of oceanic
solution by plankton, which on death settle slowly to the ocean bottom. The
calcium carbonate component consists partly of foraminifera (Table 3-2), marine

protozoa which sometimes may have arenaceous or chitinous tests (shells). In
their habit they may be shallow water, deep water or benthonic dwellers.
Benthonic (benthic) foraminifera are bottom dwellers which occur everwhere,
provided that free oxygen is available.

TABLE 3-2   Common species of foraminifera grouped according
to temperature preferences (after Phleger *et al*
1953)

Species abundant in low latitudes

Globigerina eggeri
Globigerinoides sacculifera
Globorotalia menardii menardii
Globorotalia menardii tumida
Pulleniatina obliquiloculata

Species abundant in mid-latitudes

Globigerina bulloides
Globigerina inflata
Globorotalia hirsuta
Globorotalia scitula
Globorotalia truncatulinoides

Species abundant in high latitudes

Globigerina pachyderma

Species abundant in low and mid-latitudes

Globigerinella aequilateralis
Globigerinita glutinata
Globigerinoides conglobata
Globigerinoides rubra
Orbulina univera

With a pattern of outward-flowing cold bottom water, the Atlantic is richer
in calcareous oozes than the Pacific. The increased carbon dioxide content
of the Pacific increases the solution activity of that ocean so that much
calcareous sediment, including foraminifera, is dissolved. Thus most
calcareous sediment on the floor of the Pacific is dissolved, especially
below depths of 3500 m (Berger 1970a 1970b). The level at which the
solutional rate increases rapidly is the 'compensation depth', or 'calcite
compensation depth', or 'lysocline'. Below this calcareous organisms are
less well preserved, with selective preservation posing great difficulty for
the interpretation of oxygen isotope data. As may be inferred from the above,
the 'compensation depth' is found at greater depths in the Atlantic - 5300 m
according to Arrhenius (1963). One result of this is that clays tend to
occur at shallower depths in the Pacific Ocean. Also included in the calcium
carbonate component are the coccolithoporids (unicellular algae with small
calcite platelets called coccoliths), pteropod shells, and nannoplankton
(below 60 microns in size). Coccoliths, with their special light requirements,

are confined to surface waters.

The siliceous biogenic component consists of diatoms and radiolarians.
Diatoms are found in high latitudes, especially in colder water where large
upwelling form depth occurs : for example, 60 degrees south or so in the
Atlantic, and also in the northern Pacific Ocean. Radiolaria on the other
hand are mostly found in warm water : a belt of radiolarian ooze lies north
of the equator in the Pacific Ocean.

OXYGEN ISOTOPE ANALYSIS

Evidence of climatic change has been revealed through oxygen isotope analyses
of marine fossils recovered from deep-sea cores. Although the technique may
be carried out on a variety of organisms representing different environments
(Table 3-3), the most extensively utilised are planktonic and benthonic
(benthic) foraminifera. Coccoliths and siliceous organisms (e.g. diatoms)
may also be used, but hitherto data is sparse from deep-sea cores. Studies
of corals have not been altogether satisfactory, but in some cases they are
useful indicators of former temperatures and water depth.

TABLE 3-3   Some Oxygen Isotope Studies on Quaternary
            samples

Samples	Age	Source
Carbonates		
Foraminifera	Quaternary	Emiliani 1955
Foraminifera and Pteropods	Quaternary	Deuser and Degens 1969
Corals	Holocene (recent)	Weber 1973
Molluscs	Quaternary	Epstein and Lowenstam 1953
Brachiopods	Quaternary	Lowenstam 1961
Fish Otoliths	Quaternary	Devereux 1967
Silicates		
Diatoms	Holocene (recent)	Labeyrie 1974
Radiolarians	Quaternary	Mopper and Garlick 1968
Sulphates		
Molluscs	Quaternary	Cortecci and Longinelli 1973
Phosphates		
Molluscs	Quaternary	Longinelli and Nuti 1973a
Fish Bones and Teeth	Holocene (recent)	Longinelli and Nuti 1973b
Spelothems	Holocene (recent)	Hendy 1969

It is to Harold Urey of the University of Chicago that credit must be given
for the method. He argued that, although supposedly identical in chemical
composition, the isotopes of an element did not behave similarly in chemical
processes. When water evaporates ^{16}O, ^{17}O and ^{18}O (Table 3-4) go off at
different rates. More of ^{16}O, the lighter isotope, is carried by the vapour
than that of the heavier ones. This means that the water will become enriched
in the heavier isotopes. Thus the isotopic composition of water is by no
means constant.

TABLE 3-4   Stable oxygen isotope abundances in air
            (Nier 1950)

Mass Number	% Abundance
16	99.759
17	0.0374
18	0.2039

Reasoning geologically from the above, it was a short step to the realisation
that any oxygen bearing substances precipitated from water of a given isotopic
composition, for example an organic or inorganic carbonate, should have an
identical isotopic ratio. Any change in the isotopic composition of the water
should be monitored directly by any contemporary precipitates. Urey (1971)
calculated that when oxygen isotopes are precipitated slowly in a solution of
water, fractionation takes place between water and the carbonate ions. The
fractionation factor is temperature dependent : for example, the
$^{18}O/^{16}O$ ratio changes by 0.0023 per degree centigrade. Reasoning geologically
this means that a comparison of the ratio between fossil and modern organisms
allows an estimate of the temperature at the time of life for the fossil ones
to be established. In Urey's words : 'I suddenly found myself with a
geologic thermometer in my hands'.

His initial premises were confirmed. Molluscs grown experimentally under
controlled temperature and isotopic conditions, at the Scripps Institute of
Oceanography, secreted $CaCO3$ in isotopic composition of the carbonate
secreted was temperature dependent. This yielded an empirical palaeotemperature equation derived by Epstein, Buchsbaum, Lowenstam and Urey (1953).
When used to predict the temperature of growth of recent foraminifera from
their isotopic composition, the results checked with the temperature of the
ocean water in which they grew.

It is, however, worth recalling that some time elapsed between the initial
theoretical stages and subsequent successful execution and verification of
the method. This lapse of time also helps to emphasis the delicacy of the
measurements involved (Table 3-4), for in 1947 the available instrumentation
was unable to measure the minute differences in the oxygen isotope ratios.
It was necessary to improve the sensitivity of available mass spectrometers

by about ten-fold. Nowadays the procedure outlined by Hecht (1976), with variations adopted by different authors, is normally followed (Table 3-5).

TABLE 3-5 Procedure in oxygen isotope analysis (based on Hecht 1976)

(1)   Foraminifera cleaned using ultrasonic vibrator.

(2)   Washing and sieved through a 200 micron mesh.

(3)   Oven or room drying at 50 degrees centigrade.

(4)   Microscopic scan to sort into species classes.

(5)   Sample selected : less than 1 mg. Formerly it was necessary to use samples of 3-5 mg which corresponded to about 50 to 100 individual specimens.

(6)   Sample crushed lightly.

(7)   Ultrasonic cleaning.

(8)   Washing and drying.

(9)   Either clorox may be used to remove any remaining organic matter, followed by repeated washing or samples may be roasted at 450°C for 30 minutes.

(10) Sample reacted with 100% phosphoric acid at 25°C or 50°C (different workers) for 12 to 24 hours.

(11) The gas thus liberated is purified of water vapour and collected.

(12) Analysis by mass spectrometer. Precision between different laboratories averages about $\pm$ 0.05 per mil

Table 3-5 enables an appreciation of the procedure which, perhaps, is too often taken for granted by users of the method. Results are published using the δ notation. δ is defined by :

TABLE 3-6

$$\delta = \frac{{}^{18}O / {}^{16}O \text{ sample}}{{}^{18}O / {}^{16}O \text{ standard}} - 1$$

From Table 3-6 it may be seen that measurements are expressed in parts per thousands, (parts per mil) for example : 0.5 $^O/oo$. The standard is usually

that of the PDB-1 reference, which is the isotopic value of a belemnite
rostrum from the Pee Dee Formation (Cretaceous) of South Carolina, U.S.A.
That is, a value of 0.5 per mil to PDB shows that the carbon dioxide
derived from a sample is 0.5 per mil relative to the carbon dioxide derived
from PDB-1. When the isotopic composition of water samples, including ice
and snow is reported, the reference datum is that of 'Standard
Mean Ocean Water' (SMOW) (e.g. Craig 1961. Dansgaard, Johnsen, Clausen and
Gundestrup 1973). PDB is + 0.2 per mil relative to SMOW.

Sources of uncertainty

Before proceeding to consider the results of oxygen isotope analyses it is
desirable to be aware of six major sources of uncertainty : (1) nature of
the sample analysed, (2) depth stratification of foraminifera, (3) biological isotopic fractionation, (4) selective solution of species, (5) the
role of Pleistocene ice-sheets in their effect on the changing isotopic
composition of the oceans, (6) the role of sediment mixing (bioturbation).

(1) An average sample of foraminifera may consist of several species. Even
if a monospecific sample is considered it is more than likely that its
individuals represent a time span of one or more thousand years. The
average growth period for an individual foraminifer has been estimated at
some 30 to 40 days (Berger 1967), so that its isotopic composition represents
conditions typical of a fraction of the year. Moreover, some species appear
to grow during the warmer months, and others during the cooler ones. Samples
from ocean towns (samplers) for example have shown that *Globigerinoides rubra*
grows during the warmer months, whereas *Globigerina bulloides* grows during
the cooler ones (Tolderlund and Be 1971). *Globigernoides sacculifera*, the
species most frequently used tends to dominate during the late summer and
early autumn. It may be concluded from the foregoing, therefore, that the
isotopic measurements on the tests of foraminifera represent an average
value for possibly thousands of years, and possibly for particular seasons.

(2) Most foraminifera inhabit depths to about 500 m, and most calcite
secretion   takes place in near surface waters (+ 150 m). Emiliani's (1955)
analyses show that different species record different palaeotemperatures
(Fig. 3-2), which he argued were a function of depth stratification. Surface
water foraminifera such as *Globigerinoides rubra* and *Globigerinoides
sacculifera* record palaeotemperatures of greater amplitude and detail than do
deeper water species such as *Globorotalia menardii* (100-150 m) and
Globorotalia truncatulinoides (< 200 m).

Depth stratification has long been know to occur (e.g. Rhumbler 1901), but
isotopic data has shown that it occurred throughout the Cenozoic. Not only
does it occur for single species, but also for phenotypes of the same species
(Hecht and Savin 1972). Some species even follow depth migration during
their life cycle.

According to Emiliani (1954) depth stratification is due to temperature and
water density effects. Savin and Stehli (1974), however, argued that the
basic control was osmotic equilibrium between the test and sea water; but
Shackleton (1977a) thought that latter to be inherently unlikely, and
believed that constant water density was the prime dominant. If osmotic
equilibrium is the control, it means that migration of planktonic foraminifera
down the water column occurs during glacials, and up it during interglacials,
as water is alternately abstracted and returned to the oceans as a result of

glaciation and deglaciation on the continents. Isotopic temperatures greater than those actually taking place would be registered by the foraminiferal tests. If water density is the control on depth stratification, the opposite effects would obtain. Migration would occur down the water column during an interglacial as the oceans become less saline due to the return of water from the continents. During glaciation, however, migration would take place up the water column as the oceans become more saline. Isotopic temperatures registered by the tests would be less than the actual fluctuations of the water column.

(3) Some species appear to secrete calcium carbonate in conditions other than those of isotopic equilibrium with the waters they inhabit. Such apparent biological isotopic fractionation can be referred to as the 'species effect' (Hecht 1976). It has been shown, for example, that isotopic temperatures from some present day and recent planktonic foraminifera are higher than those of the waters they live in. This is true for *Globigerinoides sacculifera* and *Globigerinoides rubra*, although in the case of the latter it may be due to the presence of symbiotic algae capable of occupying up to 80% of its shell.

Bethonic species have also yeilded isotopic temperatures which differ from the actual ones of the water they inhabit (Duplessy, Lalou and Vinot 1970); in the case of larger specimens greater deviations are recorded (Vinot and Duplessy 1973). *Uvigerina* sp. does, however, appear to grow in isotopic equilibrium with ocean water (Shackleton 1974), and hitherto has yielded the best and most consistent data (Shackleton 1977a). Because of such uncertainty, Imbrie, Van Donk and Kipp (1973) have suggested that it may eventually be desirable to establish an empirical temperature scale for individual species over a range of both test size and isotopic temperatures.

(4) Selective solution on the sea floor (Berger 1968) results in unrepresentative isotopic temperatures. The intensity of solution increased during interglacials in the Pacific Ocean, but during glacials in the Atlantic (Gardner 1975. Luz and Shackleton 1975). This means that isotopic temperatures form the Pacific are higher than those actually experienced, whereas the opposite is true for the Atlantic Ocean.

(5) Much debate has obtained concerning the effect of alternate glaciation and deglaciation on the isotopic composition of the oceans. During a glaciation water evaporated from the oceans, and precipitated as rain or snow to nourish ice-sheets, is isotopically light. That is, consisting largely of the lighter oxygen isotope, ^{16}O, which is preferentially evaporated from the oceans. Concomitant with this the oceans are residually enriched in the heavier isotope, ^{18}O, and become isotopically more positive.

The isotopic composition of foraminifera depends on two variables. One is the temperature of the waters they inhabit, the other, the isotopic composition of those waters. Unlike studies on present day and recent species, with the isotopic composition of the oceans known, uncertainty is introduced with fossil species because the isotopic composition of their contemporary ocean waters is unknown. Emiliani (1955) attempted to estimate the former composition of sea water. He calculated the maximum extent and isotopic composition of Pleistocene ice-sheets and their effect on the oceans. This amounted to a difference of 0.5 per mil between a glacial and an interglacial. Thus, because the total isotopic variation measured by him (Fig. 3-2) between

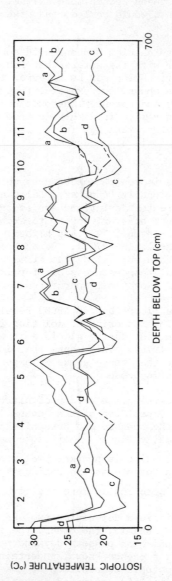

Fig. 3-2. Emiliani's (1955) palaeotemperature curves from Core A179-4 (16° 36'N, 74° 48'W). Stages 1 to 13. a : *Globigerinoides rubra*. b : *G. sacculifera*. c : *Globigerina dubia*. d : *Globorotalia menardii*

a glacial and interglacial amounted to about 2 per mil, he attributed more
than 70% of that difference to the effect of temperature change rather than
isotopic change in the composition of ocean water.

Emiliani's value of 0.4 per mil was subsequently challenged by Olausson (1965)
who estimated that value to be 0.7 per mil; Dansgaard and Tauber (1969)
estimated it as 1.2 per mil, while Shackleton (1967) argued that the value lay
between 1.4 per mil and 1.6 per mil. Thus these authorities, unlike Emiliani,
attributed most of the change recorded by fossil foraminifera to the changing
isotopic composition of the oceans rather than to temperature variation. It
follows that the major control on the isotopic composition of the oceans is
the amount of ice stored on the continents; the oygen isotope data constitute
palaeoglaciation rather than palaeotemperature curves.

Emiliani and Shackleton (1974) reasoned that with a theoretical fall in sea
level of 120 m a range of possible values could be established for the
isotopic composition of sea water during glacials. Two assumptions were made,
which give values of : (i) + 0.4 per mil if the composition of the water
evaporated was that of water vapour in isotopic equilibrium with the
subtropical sea surface, which is -11 per mil; (ii) + 1.6 per mil if the
composition of the water has the same composition as shown in central
Antarctica, which is -50 per mil. These are extreme values so that the
natural situation occurs somewhere between. Four independent means of assess-
ing the factor are possible.

(a) Table 3-7 shows estimates based on calculations of ice thicknesses,
     ice extent, and its isotopic composition, as well as changes in the
     volume of sea water, between glacials and interglacials. They range
     from 0.5 per mil, which is Emiliani's recalculated value (below)
     (Emiliani 1971) to 1.7 per mil, the median value being 1.1 per
     mil.

     Olausson (1965) esimated the isotopic composition of Pleistocene
     snow. Dansgaard and Tauber (1969) based their value on the oxygen
     isotope composition of modern precipitation and that of last
     glaciation ice recovered from the Camp Century borehole in Greenland
     (Chapter 8). Their data suggest that the isotopic composition of
     ice stored on the continents was between -30 and -35 per mil,
     which would lead to a change of ca. 1.1 per mil in the composition
     of the oceans.

(b) Shackleton (1967) argued that because the temperature of deep ocean
     water is controlled by that of Antarctic Bottom Water, it is unlikely
     to have changed appreciably during the Pleistocene. It follows that
     isotopic measurements on benthonic foraminifera indicate variations
     in the isotopic composition of ocean water alone there being no
     temperature component to take into account. The range of values so
     determined between glacials and interglacials lies between 1.1 per
     mil and 1.6 per mil for the Equatorial Atlantic and Pacific Oceans.
     Much depends, however, on the assumption that the temperature of
     botton waters has remained more or less constant (Emiliani and
     Shackleton 1974). Data which suggest otherwise are changes in test
     sizes of benthonic foraminifera from the Pacific (Shackleton and
     Opdyke 1973), and the Atlantic, where changes in species abundance
     also occurs (Streeter 1973).

(c)   Emiliani (1971) evaluated the amplitude of glacial/interglacial
      temperature change in low latitudes from data on reconstructed
      snowlines, pollen analysis, and frequency changes in the abundance
      of various foraminifera. He concluded that the change amounted to
      7.5 degrees C in the Caribbean, 5.5 degrees C in the Equatorial
      Atlantic, and 3.5 degrees C in the Equatorial Pacific. By comparing
      these data with isotopic values from fossil foraminifera he was
      able to calculate the effect of continental ice on the isotopic
      composition of the oceans as 0.5 per mil.

(d)   Transfer functions are equations which relate parameters established
      for assemblages of recent foraminifera, e.g. summer temperature,
      winter temperature, salinity, to fossil populations (Chapter 6).
      They have yielded data which shows the effect of continental ice
      sheets on the oceans to amount to 1.8 per mil, according to Imbrie,
      Kipp and Van Donk 1973, but 0.3 per mil according to Hecht (1973).

TABLE 3-7   Estimates of the effect of Pleistocene ice sheets
            on the isotopic composition of the oceans

Source	Water volume change x 10^6 km^3	^{18}O of ice (per mil)	Effect on oceans ^{18}O per mil (PDB-1)
Emiliani 1955	58	-15	0.5
Craig 1965	100	-17	1.5
Olausson 1965	65	-35	1.7
Shackleton 1967			ca. 1.7*
Dansgaard and Tauner 1969	47	-30	1.2

*Shackleton calculated the effect taking into account the flow
 of ice in three dimensions. It yielded a result 'close to
 that obtained by Olausson'.

The data assembled in Table 3-8 indicates that the effect of continental ice
sheets on ocean isotopic composition has yet to be established definitively.
On the one hand Emiliani (e.g. 1971) believes that the record in planktonic
foraminifera of isotopic change is primarily a reflection of ocean surface
temperatures, whereas on the other hand there is a formidable school of thought
which maintains that it is isotopic changes in the composition of sea water that
controls the observed variations. What is important, however, is that the
isotopic record constitutes a stratigraphical framework of global significance.
The difference of opinion noted above does not in any way alter the reality of
the oxygen isotope stages. Indeed excellent correlation may be effected
between records believed to reflect isotopic changes in the composition of

ocean water only, e.g. from benthonic foraminifera, and records believed to contain a palaeotemperature component, e.g. from planktonic foraminifera (Figure 3- 3).

TABLE 3-8   Estimates of the effect of continental ice on temperature ranges

Authors	Method	Ice effect*	Caribbean Temperature	Atlantic range °C	Pacific
Median of Table 3-7	Continental ice sheets estimate	1.1	4	3	0
Emiliani & Shackleton (1974)	Benthonic forams minimum	0.5	7	6	3
	maximum	1.1	4	3	0
Emiliani (1971)	Snow line, pollen & faunal data	0.2	7.5		
		0.5		5.5.	
					3.5
Imbrie & Kipp (1973)	Transfer functions	1.8	2.2		
Hecht (1973)	Transfer functions	0.8	5.0		

| Isotopic ranges per mil** v PDB-1 (Emiliani 1966. Shackleton and Opdyke 1973) | | | 1.9 | 1.8 | 1.1 |

*   ice effect per mil v PDB-1
**   planktonic foraminifera

OXYGEN ISOTOPE STRATIGRAPHY

Oxygen isotope stages

Following Arrhenius (1952), Emiliani (1955) recognised a succession of stages based on the isotopic variations shown by the deep sea cores. They were numbered as follows : the present warm stage (Holocene) was designated stage 1, then working backwards in time (down-core), and by reference to a model of sinusoidal temperature fluctuations, cold stages were given even numbers, and warm ones odd numbers (e.g. Figs. 3-2 and 3-5). Boundaries between stages were placed at the mid-points between temperature maxima and minima. Subsequently Emiliani (1966) recognised unconformities in some of his original cores, hence modified his original scheme accordingly. This includes 16 stages.

Broecker and van Donk (1970) believed that the pattern of temperature changes was not sinusoidal, but saw-toothed in nature. This they believed indicated the gradual accumulation of ice on the continents over period averaging about 90 ka, and deglaciation in less than one tenth of that time. Such rapid

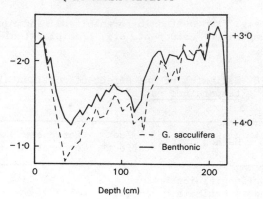

Fig. 3-3.   Isotopic analysis of
planktonic v benthonic foraminifera.
Stage 5 to 1, V28-238 (after
Shackleton and Opdyke 1973)

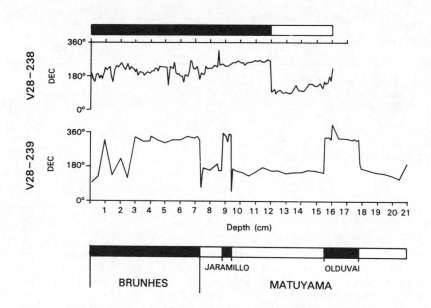

Fig. 3-4.   Palaeomagnetism of Cores V28-238 and V28-239.   Declination
plotted v depth (after Shackleton and Opdyke 1973)

deglaciations were called Terminations, and reflect sharp decreases in the ^{18}O content of the oceans. For example, Termination 1 relates to the transition from stage 2 to stage 1; and Termination 2 relates to the transition between stage 6 and stage 5*. They also recognised that there were secondary oscillations superimposed on the primary ^{18}O cycle.

Terminations, however, while conceptually useful, are based on a model of saw-toothed temperature variations which relates the growth of ice to most of the cold stage, culminating in its maximum extent fairly late. This is a matter of opinion. Furthermore, Termination, is not a stratigraphical term (Shackleton 1975), and its use in a stratigraphical context is misleading (see Chapter 4). Emiliani and Shackleton (1974) maintain the isotopic record is an actual record of worldwide events. It should not be related to any model whether it be sinusoidal, sawtooth or square wave with added noise.

A different approach was followed by Shackleton and Opdyke (1973) who defined stages at given depths in core V28-238 (Fig. 3-5). Correlation of particular units at specific localities could then be made with stages recognised in that named deep-sea core : e.g. a glacial lithostratigraphic unit could be correlated with 'the latter part of stage 6 in core V28-238'. Furthermore, they proposed that the stages they recognised be adopted as the standard for the latter half of the Pleistocene.

Dating of deep sea core stages

A means of dating stage boundaries, for at least key cores, is essential, for one isotopic cycle looks much like another, hence lacunae in some cores might pass undetected. Dates at given horizons at least allow the detailed isotopic signature between them to be characterised for comparative purposes.

Three principal means of dating deep-sea core stages are available : (1) Uranium Series dates on ocean sediments, (2) carbon-14 dates for upper core sediments, and (3) identification of geomagnetic reversals in the core sediments. Intrinsically all three methods have limitations (Chapter 5). Radiocarbon and magnetic dating of ocean cores depends upon extrapolation of their dates, usually down-core for the former, and up-core for the latter, thus relying on an assumption of constant rates of sedimentation for any validity. Uranium Series dating is now regarded with more uncertainty (Kaufman et al 1971), than when it was first proposed and utilised. Supplementary means of dating include a limited number of faunal extinctions that appear to be synchronous globally (Chapter 6), and some widely distributed tephra (volcanic ash) layers (Chapters 5 and 8).

Uranium Series datings were performed as long ago as 1942. Piggott and Urry (1942) used excess ^{230}Th to determine the age of cores from the Caribbean. Interestingly enough the date of what is now recognised as the thermal maximum stage 5 (5e), was shown to be about 100 ka. More recent work on

---

* It should be noted that stage 3, although numbered so as to indicate a warm period, is not regarded as an interglacial as is stage 1, and other off numbered stages or parts of them: hence Termination 2 does not relate to stage 3.

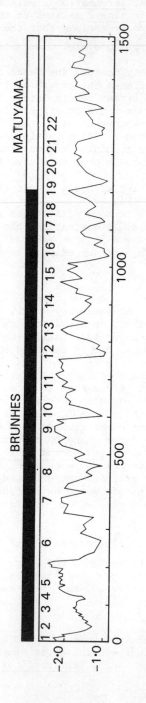

Fig. 3-5. Core V-28-238 from the Solomon Plateau (3120 m depth), Pacific Ocean (Shackleton and Opdyke 1973). Oxygen isotope stages 1-22. Core depth (base) in cm. δ deviation, per mil, from Emiliani B1 standard. This core has been proposed as a standard for the later Pleistocene. Brunhes-Matuyama reversal (700,000 years BP) occurs in stage 19 and is used as the basis for dating stage boundaries (Table 3-9)

cores from the Caribbean and Atlantic Oceans using the ^{230}Th/^{231}Pa method
gave dates of about 100 ka for the peak of stage 5 (Rosholt *et al* 1961). But
some of the dates were some 25% higher, giving 125 ka (Rona and Emiliani 1969,
Broecker *et al* 1968). It can be concluded that while such methods yield
dates that are approximately correct, sufficient uncertainties about the decay
rates of daughter isotopes remain to render them at least partly equivocal
(Ku *et al* 1972).

Radiocarbon dating was applied to deep-sea cores by Robin and Suess (1955)
and has since proved a valuable measns of dating as far back as that method
allows (Chapter 5). Beyond that ages need to be extrapolated, a not
altogether satisfactory situation because evidence exists for systematic
long-term trends in rates of sedimentation in, for example, the Caribbean
(Emiliani and Shackleton 1974). In that region, at least, extrapolation
of radiocarbon dates must be regarded with caution.

The application of palaeomagnetic stratigraphy (Hays *et al* 1969)(Chapter 5) has in
many ways revolutionized the study of deep-sea cores in that well-founded
dates may now be interpolated at various points. Three deep-sea cores in
particular have been investigated in detail from the point of view of their
palaeomagnetic and oxygen isotope stratigraphy. In other words placing the
oxygen isotope stratigraphy in a palaeomagnetic time-scale. Two of them
extend to the base of the Pleistocene and beyond into the Pliocene. These
are cores V28-239 (Fig. 3-4) from the Pacific Ocean (Shackleton and Opdyke
1976), and core V16-205 (Fig. 3- 6) from the Atlantic Ocean (van Donk 1976).
The location of core V28-239 is adjacent to that of vore V28-238 (Shackleton
and Opdyke 1973), which at that time was the only one to be studied from both
oxygen isotope and palaeomagnetic standpoints. V28-238 contains the Brunhes-
Matuyama magnetic boundary, age 700 ka, at a depth of 1200 cm in the core, and
within stage 19 (Fig. 3- 5). By assuming constant sedimentation (1.74 cm/10^3
years) the boundaries of the 22 stages defined were dated (Table 3-8). Stages
1 to 16 correspond with those errected by Emiliani (1966), and all 22 are
defined at particular depths in the core (Table 3-8). The peak of stage 5,
5e, was estimated as 123 ka. This corresponds closely with a large number of
Uranium dates on raised beaches and raised coral reefs throughout the world
which range from 120 to 125 ka. These represent the high stand of sea level
during the last interglacial (Chapter 7) which Shackleton (1969) had
previously argued corresponded not to all of stage 5, but merely to 5e. This
opinion is vindicated not only by data from Caribbean core P6304-8 which shows
that the sea was cooler during the later part of stage 5 (Emiliani 1966), but
also by data from Barbados which demonstrates a fall in sea level 110 ka ago
of fully glacial proportions (Steinen *et al* 1973): that is, immediately
following stage 5e (Chapter 7). Stage 5e has also been dated as 120 ka by
down-core extrapolation, assuming that the temperature minimum of stage 2
(at 35 cm depth in V28-238) is 20,000 radiocarbon years.

In an evaluation of the dates they proposed for stage boundaries in core
V28-238, Shackleton and Opdyke (1976) concluded they were still the best
available, although they ought not to be regarded as in any way definitive.
Future work is almost certainly going to modify them. In passing it is worth
noting that their estimates (Table 3-8) are somewhat greater than those
proposed by Broecker and van Donk (1970), and those earlier by Emiliani (1966).

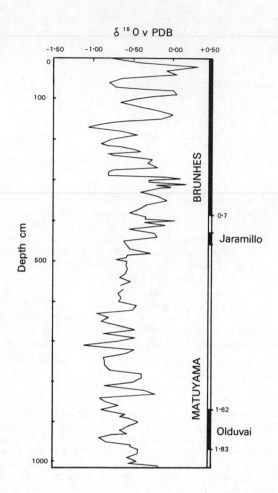

Fig. 3-6.   Oxygen isotope analysis and palaeomagnetism of Core V16-205 for the entire
Quaternary.   Polarity time scale in Ma. Jaramillo :
0.89-0.95 Ma (after van Donk 1976)

Of the 22 stages defined by Shackleton and Opdyke (1973), stages 1 to 18
occur in the Brunhes (magnetic) Epoch. Because stages 2, 3 and 4 are
recognised as corresponding to the last glaciation, as strictly speaking so
is most of stage 5 (with the exception, that is, of 5e, which is the last
interglacial),this means that eight major glaciations of Brunhes age occurred :
stages,18, 16, 14, 12, 10, 8, 6 and the last glaciation (2-3-4-5a-5b-5c-5d).
Somewhat earlier, stage 22 is recognised as the first glacial event which
corresponds in magnitude to those of the Brunhes Epoch. It would seem likely,
then, that the complete glacial-interglacial sequence of, for example, Europe,
occurred in the last 800 ka, a proposition corroborated by the available data
from that continent (Chapters 6 and 8). Farther back, glacial fluctuations
continue to the Olduvai magnetic event, thence with less regularity into the
Pliocene.

While at first sight there may appear to be cyclicity displayed by the record,
with glacial cycles lasting about 100 ka and increasing in amplitude toward
the more recent stages, this is more apparent than real. Every minimum, with
the exception of stage 14, reaches more or less the same isotopic value.
Maximum isotopic values are much the same, with the exception of stages 3 and
17, which are lower, and stages 1 and 5, which are higher. Stage 5 is the
highest, a feature of considerable significance for changes in sea level
(Chapter 7). The onset of each interglacial is rapid (Broecker and van Donk
1970), a feature confirmed by quantitative analyses on marine fossil populat-
tions (Chapter 6). For example, the isotopic data shows deglaciation about
12 ka ago to have been so rapid, that its record is as much a function of
sediment mixing as actual isotopic change, being some 0.3 per mil/
10^3 years. The same is true for deglaciation and the onset of interglacial
conditions at the end of stages 2, 6, 10, 12 and 22; but not at the end of
stages 4, 8 and 14.

It is not possible at the present time to make any further generalizations.
Nor is it possible to characterize precisely any given stage in climatic
terms. No translation of the isotopic values into climatic parameters at any
given locality is possible. Hence the persisting difficulty in correlating
continental climatic episodes with oxygen isotope stages in the oceans.
This difficulty has already been touched upon with regard to the last interglacial. According to the north-west European definition of an interglacial
it corresponds to a period characterised at its thermal peak by the presence
of mixed oak deciduous forest woodland. Shackleton (1969) has shown, however,
that the last such event in Europe corresponds to stage 5e. The rest of
stage 5 consisted of colder climates - even colder than he originally
estimated (Shackleton and Opdyke 1973). If the duration of the last interglacial (5e) is typical of earlier interglacials, it means that they constitute only some 10% of the time. Yet, the marine isotopic record shows that
cold and warm stages are of roughly equal duration. Clearly the oxygen
isotope record cannot resolve such problems. Questions of definition are
involved. How should an interglacial be defined? Is the palaeobotantical
definition used in Europe too limited and too parochial in scope? Instead
should definition be in global terms? These are matters outside the scope of
this chapter and are deferred until later.

It is as a common technique of global applicability that oxygen isotope
stratigraphy achieves its greatest value and potential. Enough independent
criteria are available to demonstrate that inter-core correlation, to a high
degree of detail and accuracy, is possible throughout the world's oceans.

Oxygen isotope stages are synchronous globally with an accuracy only limited by the mixing time necessary for isotopic changes initiated in one part of the world's oceans to be transmitted everywhere. Certainly for the more recent stages this is less than 1,000 years, thus providing isochronous surfaces of hitherto unheard of accuracy in the whole of the stratigraphical column. By way of example, rapid deglaciation of the Laurentide ice sheet obtained 14 ka ago and was recorded more or less immediately in the Gulf of Mexico as cold, isotopically light, water drained along the Mississippi valley (Shackleton and Kennet 1976). It was not recorded in the Pacific Ocean for several centuries, but within 1,000 years. The finite mixing time of the ocean has also been determined from other standpoints, with various estimates, but all less than 1,000 years (Gordon 1975).

Accepting the proposition that : 'it is highly unlikely that any superior stratigraphic subdividion of the Pleistocene will ever emerge' (Shackleton and Opdyke 1973), it seems not illogical to proceed with it as a stratigraphical framework for studying the Quaternary, not only for the oceans but the continents as well. A principal goal is the correlation of continental events with the oxygen isotope stratigraphic record of the oceans. It is to that end that attention can now be directed.

TABLE 3-9   Age estimates for stage boundaries in core
            V28-238 (Shackleton and Opdyke 1973)

Boundary	Depth (cm)	Age (yr.)
1-2	22	13,000
2-3	55	32,000
3-4	110	64,000
4-5	128	75,000
5-6	220	128,000
6-7	335	195,000
7-8	430	251,000
8-9	510	297,000
9-10	595	347,000
10-11	630	367,000
11-12	755	440,000
12-13	810	472,000
13-14	860	502,000
14-15	930	542,000
15-16	1015	592,000
16-17	1075	627,000
17-18	1110	637,000
18-19	1180	688,000
19-20	1210	706,000
20-21	1250	729,000
21-22	1340	782,000

Chapter 4
CLASSIFICATION

When assessed against the comparative perfection of Pleistocene events as recorded in deep-sea cores (Chapter 3), the continental record is meagre, and to various degrees seemingly intractable (Chapter 2). Yet, inadequate though earlier inter-regional and inter-continental correlations now appear to be, the global nature of the ocean record, which is without parallel in pre-Quaternary rocks, presently holds out an objective yardstick, or potential standard stratigraphic scale, for ultimate correlation. Indeed there is every reason why Shackleton and Opdyke's (1973) proposal, that the record of deep-sea core V28-238 be taken as standard for the later Pleistocene, should be accepted, if only provisionally. Eventually it is likely that similar standards for the earlier Pleistocene will be proposed. Distant though any realization of continental and oceanic correlation may be, it is worth noting that some particularly favoured continental data, for example, the loess sequences of central Europe, already allow convincing correlation with the deep-sea record (Chapter 9) (Kukla 1970. 1975).

Before lasting progress can be made it will be necessary to clarify much of the confusion that presently attends continental stratigraphy. The stratigraphic reassessment of Deevey (1965) is slow to make its mark. Some of this will merely involve the restatement, or barely modified upgrading, of soundly based earlier work : for example, as by the Subcommission on North American Stratigraphy of INQUA, 1977. Less straightforward will be reclassification of stratigraphic data, presently ambiguous, perhaps founded on redundant concepts, inadequately defined, or the result of misconceived stratigraphic status. But, how should Quaternary rocks be classified?

It could be said that three principal schools of thought exist. The first sees no reason for regarding the Quaternary Period as being in any way fundamentally different from pre-Quaternary periods in that similar rocks were deposited. Hence classification should be the same for both. Here belongs the Stratigraphic Commission of INQUA, who have adopted the most recent stratigraphic guide of the International Union of Geological Sciences (Hedberg 1976) as the basis for classification of Quaternary lithostratotypes (type sections for lithological units).

The second would argue that there are certain distinguishing features that make the Quaternary Period and its rocks a special case (below). That being the case, then, special rules should apply. These are usually a mixture of stratigraphic and other comparative criteria. They include geomorphology and periglacial criteria as a means of classification, as well as the extension of otherwise soundly based stratigraphic procedures beyond the limits of their usefulness and resolving capability. The latter can, in some instances, include the use of assemblage floras or faunas, as a means of dating (e.g. as advocated by Luttig *et al* 1969 in the case of pollen analysis). In its reliance on basic stratigraphic method, however, it ought not to deviate from procedures recommended by various codes. It could be argued that when such departures are made then the greatest confusion usually obtains. The use of landforms is a particularly good case in point (Chapter 8).

The third school of thought undoubtedly views stratigraphic classification as something of a hindrance, and instead advocate the use of radiometric dates as a means of subdivision, correlation and dating (e.g. Vita Finzi 1973). This rests on a unrealistic faith in such dating procedures. So many potential pitfalls and errors are inherent to existing methods (Chapter 5) that the whole field may be likened to the minute tip of an enormous experimental iceberg. Such dating is hardly suited to the task demanded of it at the present time.

THE QUATERNARY AND PRE-QUATERNARY COMPARED

Persistent among the claims for treating the Quaternary somewhat differently from other periods are those which point to its short time-span, inadequacy of its palaeontology for conventional utilization in subdivision, fragmentary nature of its depositional record, unusual stratigraphic relations caused by the influence of geomorphic situation, and the way in which it has been subdivided using climate as the standard.

Several estimates exist as to the length of the Quaternary. They range from 0.6 Ma (Emiliani 1955) to greater than 4 Ma (Savage and Curtis 1967). Now that earlier Cenozoic glaciation is an established fact in many parts of the world attempts to use Pleistocene and Ice Age(s) as synonyms are invalid. Based on multiple criteria, mostly of a marine nature, which is to be encouraged in the definition of a major standard chronostratigraphic unit such as a Period, a recent estimate of the base of the Quaternary as 1.6 Ma (Haq, Berggren and Van Couvering 1977) seems desirable for another reason also. For it coincides with a geomagnetic reversal (top of the Olduvai Event) which is recognised on a global basis (Chapter 5).

Given the brief duration, therefore, it means that Quaternary scientists are concerned with time-spans of a few thousand of years, sometimes of hundreds of years - especially within the range of radiocarbon dating, and only occasionally with spans of, say, 0.5 Ma. This is in stark contrast to pre-Quaternary time-spans, which as working units, commonly amount to several millions of years. Thus available Quaternary time planes are infinitely more finely tuned than are earlier ones. Proximity to the present renders a degree of detail available from most evidence unparalleled from earlier deposits. On this basis alone a shift in emphasis, if not in basic philosophy, could be pressed. There are few who make significant contributions to Quaternary as well as pre-Quaternary geology. A remarkable characteristic of Quaternary is that many of its fossil groups have modern counterparts that are anatomically identical. Where evolutionary changes are demonstrable they are usually of limited value for subdivision purposes. By and large the brevity of the Quaternary did not allow wholesale evolution in all fossil groups. Significant evolution did, of course, occur. Hominids are an outstanding example (Chapter 6); but they are of limited value as a means of subdivision. Evolution in rodent faunas of Europe promises to become a tool of great usefulness, but has yet to fulfil its potential. Generally, when significant subdivision is possible, as in the case of North American mammalian faunas, the resulting divisions are too broad to be of much use. Three 'Mammal Ages' have been recognised in North America : Blancan (which overlaps with Pliocene) and Irvingtonian, which together span practically all of the Pleistocene, and Rancholabrean (Chapter 6).

* recent periods, easier to reconstruct.

Palaeontology then, the primary criterion for subdividing pre-Quaternary rocks, attains only limited application. Instead the fossil record attains maximum usefulness from a palaeoecological standpoint. Assemblage faunas and floras, repetitive in time and space, allow valuable inferences to be drawn about changing environments. But their inherently time-transgressive (diachronous) nature renders them unsuitable for correlation and dating at other than strictly local levels.

Added to the difficulties already mentioned is the fact that the preserved record of Quaternary events consists of a fragmentary and discontinous body of deposits to say the very least. This is generally true, though particularly so within, and immediately outside, formerly glaciated regions. To some extent, therefore, it is perhaps unfortunate that most existing classifications were formulated in such regions. The repetitive advances of glacier ice, fluvial erosion in periglacial conditions, and mass movement of slopes, combined to erase most effectively the depositional record of earlier events. Sequences of continuously deposited strata are exceptional. Notable amongst these, however, are the loess terrains of central Europe (Chapter 9), and deposits preserved in lakes or deep sedimentary basins - for example in Bogota, Colombia (van der Hammen 1964), Macedonia (Wijmstra 1969), or Japan (Horrie 1976). These are of crucial importance in calibrating the continental record, and in provisional correlating it with that of the deep oceans.

Due to inadequacy of the fossil record, and because the earliest classifications were formulated in areas formerly glaciated, the Quaternary has, and is, subdivided on the basis of climatic change. Glacials and interglacials are recognised and are usually accorded the status of stages (below). Further subdivision of the glacials recognises interstadials, which are episodes of climatic amelioration during an otherwise cold period. Some classifications also recognise sub-stages.

Classification on the basis of climatic change distinguishes the Quaternary from earlier periods. It gives rise to special problems. Classification is based not so much on the rocks themselves, but on climatic inferences drawn from them. Great care must be taken to distinguish between rock data and climatic inferences arising from them. Sometimes this is complicated by the ambiguous nature of some Quaternary deposits. Even in glaciated regions the recognition of glacial deposits is not always straightforward : many so claimed have subsequently been shown to be periglacial in origin (Chapter 8). Even when the genesis of a deposit has been established its climatic significance is not always apparent. Take for example the evidence of high lake levels in central Africa (Chapter 9). At one time evidence of such pluvial conditions was taken as being synonomous with glaciation (Chapter 2). But now it is realized it means no such thing, and actual situations are more complicated than simple models, on a one to one cause and effect basis, would suggest.

Given a classification resting primarliy on inference, an abundance of deposits of ambiguous origin and climatic significance, then problems of definition inevitably arise. While it may be comparatively straightforward to recognise glacials and interglacials in formerly glaciated regions, how are they recognised and defined elsewhere? How and where should the boundaries be drawn? Most boundaries are based on data that are time-transgressive : for example, an interglacial recognised on the incidence of temperate vegetation, as fossil pollen grains, will be recorded earlier in mid then in high latitudes. Following climatic amelioration a time-lag will occur before

the higher latitude areas are colonised by vegetation of temperate affinity.
Marine transgression was the cause of time-transgressive entities in pre-
Quaternary periods; in the Quaternary climatic change is the cause. It may
have been a swifter process than marine transgression, but its recognition
rests on very different criteria.

Repetitive climatic cycles produced equally repetitive sequences of rock.
Although of different age, a rock sequence in one locality may look much like
another which is of greater or lesser distance away. They may be said to be
homotaxial. Homotaxis is defined as:

> Rock-stratigraphic (lithostratigraphic) or biostratigraphic
> units that have a similar order of arrangement in different
> locations but are not necessarily contemporaneous are said
> to be homotaxial (American Code 1961)

Criteria are therefore necessary for distinguishing age relationships
between such homotaxial occurrences.

It cannot be denied that the Quaternary does have problems peculiarly its own.
But they are not so great as invalidate in any way the stratigraphic procedures applied to pre-Quaternary rocks. With possible ambiguity of origin and
climatic inference from strata looming so large, it is all the more desirable
to work according to objectivity defined base lines. Special situations have
been catered for by defining new kinds of stratigraphic units. These include:
morphostratigraphy, soil stratigraphy, geologic-climate units, as well as
other kinds utilized in other rocks also, such as magnetostratigraphy.
Together they supplement the fundamental modes of classification, namely:
lithostratigraphy (rock-stratigraphy), biostratigraphy, and chronostratigraphy
(time-stratigraphy).

DATA PRESENTATION : CLIMATIC CURVES

Arising from the climatic basis of recognising cold and warm, or temperate,
intervals, it becomes customary to present data in the form of climatic
curves (e.g. Fig. 4-1). More than one variety exists. One of the most
common represents frequency or frequencies of a class of data, for example,
fossil pollen grains (Fig. 6-). Usually, though not invariably, this data
is derived from continous sequences of deposits. Another integrates
disparate items of data into apparently continous sequence. An example of
this would be a climatic curve representing the glacials and interglacials
of the British Isles (Fig. 2-14). The data from which this curve is derived
is geographical separated, and only rarely from continuous sequences where it
can be shown that one stage follows, or directly precedes another. The
potentially misleading aspects of such curves are many. Some important
episodes may be missed out. In the case of the British example it has already
been argued that evidence exists for more interglacials than the climatic
curve shows (Sutcliffe 1976). The seductive aspect of such curves should be
minimised by either drawing them as discontinuous, at points where the data
is reliable, or as broken lines throughout. Continous graphs too often harden
and become monolithic standards of reality.

Given an appreciation of the nature of such curves, and the evidence on which
they are individually based, they may be used to good effect, not only for the
presentation of data (Fig. 4-1), but also as a research tool (Fig. 4-2).
Porter (1971) used time-distance curves, calibrated by radiocarbon dating, to

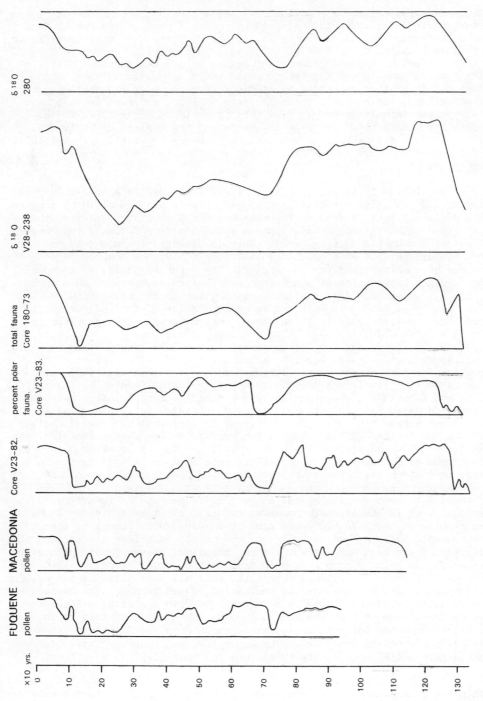

δ¹⁸O
280

δ¹⁸O
V28-238

total fauna
Core 180-73

percent polar
fauna.
Core V23-83.

Core V23-82.

MACEDONIA
pollen

FUQUENE
pollen

×10³ yrs.   0   10   20   30   40   50   60   70   80   90   100   110   120   130

Fig. 4-1. Climatic curves summarizing and correlating Upper Pleistocene data (after van der Hammen 1974)

study and characterize the position of glacial termini through time in
western North America. including the North-Central part of the Brooks Range,
Alaska, northeast St. Elias Mountains, Yukon Territory, and the Puget-Fraser
Lowland of Washington and British Columbia. This provides evidence of what
he termed first and second order fluctuations of glacier margins. First
order fluctuations lasted for ten thousand years or more, and were to be
recognised in most of the areas he studied. Second order fluctuations, on
the other hand, lasted for a thousand years or less, and were not apparent
in all regions simultaneously. From this it could be argued that these
represent the effects of local environments and the regimen of individual
glaciers. Such data should, therefore, be interpreted with care. Reliability
of the radiocarbon age control is a further factor that must be evaluated
(Chapter 5).

Very often it is only in this way that widely separated sets of data may be
compared and tentative correlation proposed. It is also possible to compare
quite different sets of data in the same way (Chapter 11). Data from the
Pacific and Atlantic Oceans was compared with data from Colombia and Greece
by van der Hammen (1974) (Fig. 4-1). This is illustrative not only of
curve comparison, but also how individual time-scales for respective sets of
data may be combined to mutual advantage. The Greek sequence, in Macedonia,
shows changes between forest and steppe from pollen analysis of 35 m of
sediment. Radiocarbon dating allowed calibration as far back as 50,000 BP.
Below that level dating is based on an extrapolation of the radiocarbon
chronology. Pollen analysis in Colombia, Lake Fuquene, was undertaken on a
core 45 m long (Wijmstra 1969). Two radiocarbon dates, the oldest being
20,575 BP allowed dating for the uppermost 11 m. Below that dating is based
on extrapolation again, taking into account rates of sedimentation established
for the upper part of the core. The marine data consists of: (1) a core
from the North Atlantic analysed by quantitative data (Chapter 6) (Sancetta,
Imbrie and Kipp 1974), where dating is estimated by interpolation between
radiocarbon dated points, extrapolation and comparison with other data; (2)
two curves representing analysis of changes in foraminiferal faunas (McIntyre,
Ruddiman and Jantzen 1972); and (3) two oxygen isotope curves (Emiliani and
Shackleton 1973. Shackleton and Opdyke 1973), the dating of which has already
been discussed (Chapter 3). Correlations such as these are discussed in more
detail in Chapter 11, but a major methodological point may be made here.

Had these data been available, and subject to comparison, a few years
previously, it is likely that the lower end of their time scales would have
been terminated about 70,000 years ago. This was, until recently, the
date conventionally agreed as being the end of the Last Interglacial and the
beginning of the Last Glaciation. In the meantime, however, data pointed to
ca. 125,000 BP being a more realistic date for the Last Interglacial (Oxygen
Isotope Stage 5e). It follows, therefore, that all such climatic curves, for
whatever data, may be stretched or contracted, so as to match the available,
or new, dating evidence. Many such curves have been replotted using dating
assumptions different from the original work. An example of this is the
modification of the Camp Century climatic curve from the Greenland Ice Sheet.
Originally plotted by Dansgaard, Johnsen, Møller and Langway (1969), it was
subsequently modified by Imbrie (Dreimanis and Raukas 1973).

Within the constraints of their radiometric tags, which are unfortunately
often equivocal, all climatic curves are elasticated to some extent. Good
quality dating control will clearly enhance their reliability, but before any

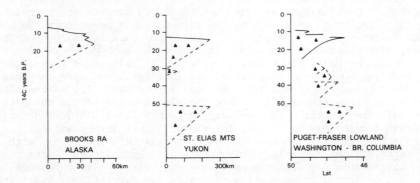

Fig. 2-2.   Time-distance graphs of glacial units
in the Western United States (after Porter 1971)

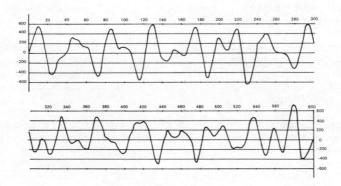

Fig. 2-3.   Changes in solar radiation (in canonic
units) for the last 600,000 years, latitude 65°N
(based   on Milankovitch, after Zeuner 1959)

are taken as standards the assumptions on which they have been drawn should
be carefully considered. Injudicious use of inadequate data all too frequently
becomes conventional wisdom.

Climatic curves used as a basis for classification should be compiled from
tangible geologic evidence, and not from theoretical considerations. While
it is certainly permissible, and even desirable, to compare and correlate
theoretical with actual data curves, the former should never form the basis
of a classification or chronology. An outstanding example of such procedure
exercising disproportionate influence is that of Milankovitch's (1941) climatic
curves. Using the rules of celestial mechanics he calculated the volumes of
solar energy reaching the atmosphere at distinct latitudes over the past
600,000 years (Fig. 4-3). His data was used as the basis for a chronology,
and geological evidence was fitted into the pigeon-holes of his theoretical
cycles. This was practised particularly by Zeuner (1945. 1959) who erected
an elaborate terminology including, for example, Last Interglacial,
Penultimate Interglacial, Antepenultimate Interglacial, and so on, with
similar terms for the glacials. This became well established in the
European literature, particularly among archaeologists. Its influence has
already been mentioned on European geologists, for Eberl (1930) modifying
the Penck and Bruckner (1909) scheme in the Alps was a supporter of
Milankovitch (Chapter 2). But it was under Zeuner's influence that the
Milankovitch model attained its greatest influence.

Recent attempts to correlate geological and theoretical climatic data on this
basis include those of Evans (1971) and Hays, Imbrie and Shackleton (1976).
Unfortunately miscorrelation at one point on his scale rendered much of
Evans' (1971) comparisons dubious. His geological data was an integration of
disparate stands of evidence of many kinds, including geomorphic features.
On the other hand Hays, Imbrie and Shackleton (1976), members of the CLIMAP
project, were able to compare the relatively objective data derived from
deep-sea cores with theoretical cycles of solar radiation input. They were
able to conclude that changes in the earth's orbital geometry were the
fundamental cause of Quaternary changes in climate (Chapter 11).

This does not mean that such theoretical climatic curves may be used as a
basis for Quaternary classification. It does not in any way alter the need
for careful and objective classification of Quaternary rocks. This will
enable workers throughout the world to understand and use more localised
data with full confidence. In particular:

> It seems important not only to state that lithostratigraphic
> units are the basis for all other types of units but that
> workers *should be seen* to proceed from actual visible rock
> units to an assessment of their lithological content and
> finally to make *inferences* as to the climatic conditions
> under which they were deposited (Bishop 1970).

STRATIGRAPHIC CLASSIFICATION

As previously mentioned there are three principal categories of stratigraphic
classification:

Lithostratigraphy (rock stratigraphy), which is concerned
with the organization of strata into units based on their
lithological characteristics

Biostratigraphy, which is concerned with the organization
of strata into units based on their fossil content

Chronostratigraphy (time-stratigraphy), which is concerned
with the organization of strata into units based on their
age relationships.

Other kinds of stratigraphic units are discussed in the following where
appropriate: for example, soil stratigraphy and morphostratigraphy, both
untilized in Quaternary investigations as basic concepts, are included in the
discussion on lithostratigraphy. Geologic-cimate units are included with the
discussion on chronostratigraphy. And magnetostratigraphy considered in
Chapter 5.

Stratotypes

In order to design a stratigraphic unit a type section (stratotype) is
necessary. This provides an objective standard on which the concept is
based. Its prime importance is that it will remain 'the permanent objective
standard available for reference, unaffected by changing bases of lithological
or biological correlation' (George 1970).

Three basic types of stratotype are recognised:

Unit stratotype : the type section which is the standard
for the definition and recognition of a stratigraphic
unit.

Boundary stratotype : a specific point in a sequence that
indicates a stratigraphic boundary.

Composite stratotype : a unit stratotype made up of the
combination of specified units of strata called component
stratotypes.

The INQUA stratigraphy commission recommend that areal stratotypes (a
particular case of the composite stratotype) should be employed. They
defined it as:

An areal stratotype includes a number of sections or other
data disposed in the area, within the limits of which the
stratigraphic range and boundaries of a named stratigraphic
unit are determined by all means including geomorphological
or facial correlation of lithologically and genetically
heterogeneous deposits (for example - a stratigraphic subdivision of glacial sequence by means of correlation of
tills, marginal moraines, glacifluvial outwashes and
adjoining river terraces).

Every stratotype has a type locality and a type area : for example see the
description of the Jessup Formation, Indiana (Table 4-1). Additional
terminology which is designed to assist understanding and to extend the
applicability of a stratotype concept are:

Holostratotype : the original stratotype of a unit or boundary

Parastratotype : supplementing definition of the original stratotype, and providing information further to that of the holostratotype, lectostratoype or neostratotype

Lectostratotype : selected subsequently if the holo-stratotype turns out to be inadequate

Neostratotype : a new stratotype to replace one destroyed or nullified

Hypostratotype : which aids the extension of a holostrato-type, lectostratotype or neostratotype into other regions.

In general the first three kinds of stratotype, holostratotypes, parastratotypes and lectostratotypes will be found in the type area. But neostratotypes, lectostratotypes and hyperstratotypes may also occur outside the original area.

The stratigraphic guide adopted by INQUA's Stratigraphic Commission (Hedberg 1976) recommends that stratotypes should be defined so as to allow ready accessibility in the field. At first sight this might appear to pose problems insofar as stratotypes designed from ocean cores are in question. But the guide makes allowance for subsurface stratotype provided that adequate samples and logs are available. Deep-sea cores are retained and curated at various institutions, hence are available for subsequent study.

Lithostratigraphy

Lithostratigraphy is the basis of all other stratigraphy hence it is partic-ularly important that is units are carefully and adequately defined. They should be defined on the basis of lithological character only, and not on any inference as to mode of origin or age. Definition should be in terms of a stratotype, or unit stratotype, wherein boundaries are usually sharply defined, though they may be placed within a transitional zone of lithological change. The overriding consideration is overall lithological similarity within the unit defined, though in the case of many glacial units, variation in lithology as between fluvioglacial sand and gravel and till, in itself constitutes a form of unifying feature.

A lithostratigraphic unit should be capable of being mapped in the field and its geometry as a tabular entity of rock established. Subsurface data assists in definition, and it is advantageous if units can be represented as cross sections. Whereas many Quaternary lithostratigraphic units have been described at a type section, their lateral and vertical variation is indeterminate, hence definition is incomplete.

Hierarchy of Lithostratigraphic Units

Formal lithostratigraphic units are defined as follows (Hedberg 1976):

Unit	Definition
Group	Two of more formations
Formation	The primary unit in lithostratigraphy
Member	A named unit within a formation
Bed	A named unit or layer in a member or formation.

Despite a recommendation that the bed should constitute the 'lithostratigraphic base unit' by Luttig, Paepe, West and Zagwijn (1969), there seems to be no good reason why the formation ought not to function as the primary working unit of Quaternary stratigraphy. Their proposal fails to meet a cardinal requirement of a basic working unit, namely that it should be capable of being mapped as a tabular body of rock strata. Beds in Quaternary stratigraphy are usually no more than local in significance, although some marker horizons, such as tephra, are of widespread importance. Their proposals, stem from the belief that pollen analysis should be used as the fundamental means of subdividing Quaternary successions on a glacial and interglacial basis, and as such is distinctly north-west European in flavour (Chapter 6).

Employing the hierarchy of lithostratigraphic units in any given area calls for a number of decisions to be made. At one time it was common for all the glacial and fluvioglacial deposits in a region to be lumped together as one formation, even if that status was not explicitly invoked. Such as procedure, however, is not likely to benefit understanding of lithologic complexity of Quaternary deposits. On the other hand it is possible that over-enthusiastic subdividion might go beyond the point which allows better understanding. No good purpose is served by subdivision for its own sake.

There are no strict rules regarding the thickness, or number, of formations that can be recognised in a region. Much depends on local circumstances and on the means best suited to understanding the lithological complexity most adequately. In Illinois, for example, Willman and Frye (1970), in subdividing the glacial, fluvioglacial and fluviolacustrine deposits, defined seven formations and thirty four members. In addition they recognised further units of formational status for various loess deposits of different ages (Table 4-2). This outstanding example of logical and objective classification, although subsequently modified in part (Johnson 1976), covers all the Quaternary deposits of that state, and they are defined carefully in terms of stratotypes and their extent indicated.

This contrasts with the large number of formations proposed for the northern part of the Isle of Man in the northern Irish Sea (Thomas 1976). There, eight formations, all of Upper Pleistocene age (cf with the number recognised for the entire Quaternary in Illinois), are recognised. None are defined in terms recommended for stratotype description by the 1976 or earlier codes. Nor is the extent of individual formations specified. Many are based on exposure in coastal sections only, and their extent inland as tabular bodies of rock strata is unknown. Recognition and proposals for lithostratigraphic units should follow recommended procedures, otherwise only limited value accrues.

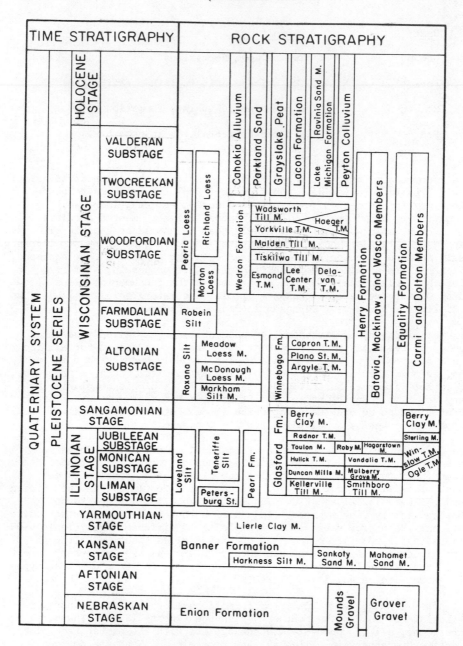

TABLE 4-2   Classification of Quaternary deposits of Illinois
using multiple criteria (from Willman and Frye,
Illinois State Geological Survey, Bulletin 94,
1970).   See also facing page.

SOIL STRATIGRAPHY	MORPHOSTRATIGRAPHY		
Modern Soil			
	Lake Border Drifts		
	Zion City D.		
	Highland Park D.		
	Blodgett D.		
	Deerfield D.		
	Park Ridge D.		
	Tinley D.		
	Valparaiso Drifts		
	Palatine D.		
	Clarendon D.		
	Fox Lake D.		
	Roselle D.		
	Westmont D.		
	Keeneyville D.		
	Cary D.		
Jules Soil	Wheaton D.		Iroquois D.
	West Chicago D.		
	Manhattan D.		
	Wilton Center D.		
	Rockdale D.		
	St. Anne D.		
Farmdale Soil	Minooka D.		
	Marseilles Drifts		
	Ransom D.	Barlina D.	Gilman D.
	Norway D.	Huntley D.	
	Cullom D.	St. Charles D.	
	Farm Ridge D.	Gilberts D.	Chatsworth D.
Pleasant Grove Soil	Mendota D.	Elburn D.	
	Arlington D.	Strawn D.	Ellis D.
Chapin Soil	Shabbona D.	Minonk D.	Paxton D.
	Paw Paw D.	Mt. Palatine D.	
Sangamon Soil	La Moille D.	Varna D.	
	Theiss D.	El Paso D.	
	Van Orin D.	Fletchers D.	
	Dover D.	Eureka D.	Illiana Drifts
	Arispie D.	Normal D.	Gifford D.
	Bloomington Drifts		Newtown D.
	Marengo D.	Metamora D.	Urbana D.
	Providence D.	Washington D.	Rantoul D.
Pike Soil	Buda D.	Kings Mill D.	Champaign D.
	Sheffield D.		Ridge Farm D.
	Shaws D.		Hildreth D.
			West Ridge D.
Yarmouth Soil			Pesotum D.
		Shirley D.	Arcola D.
		Le Roy D.	Cerro Gordo D.
		Heyworth D.	Turpin D.
			Shelbyville Drifts
	Harrisville D.		Paris D.
Afton Soil	Temperance Hill D.		Nevins D.
	Atkinson D.		Westfield D.
	Alluvial terraces are informally named in local areas		

Some Quaternary lithostratigraphic units are homogeneous for considerable
distances, others are highly variable, but provided some sort of overall
lithologic similarity is possible, then given forms of diversity can be used
in their definition. Thus, in the Lancashire-Cheshire-Shropshire lowland of
England, Worsley (1967) described the lithostratigraphic relationships for a
region that had been dominated by the *tripartite concept*. This recognised a
subdivision of the glacial deposits into Lower Boulder Clay, Middle Sands,
and Upper Boulder Clay. It had lasted for over a century (Hull 1864) and was
used as a basis for mapping by the Geological Survey of Great Britain. These
lithostratigraphic units were adopted as chronostratigraphic ones : that is,
by recognising two discrete glaciations separated by a period of retreat
(Middle Sands). Poole and Whiteman (1966) correlated the Lower and Upper
Boulder Clays with 'Würm I' and 'Würm II' of Europe respectively. Subsequently
Boulton and Worsley (1965) maintained that the tripartite scheme was untenable.
Boulton (1972) argued that many of the sequences could be explained as the
result of one glaciation only. By adopting the American Code (1961) Worsley
(1967) recognised a Stockport Formation, which included tills and fluvio-
glacial sands, separate entities of which were accorded member status (Fig.
4-4). All these were deposited during one advance and retreat of a lobe of an
Irish Sea ice sheet.

This example is illustrative of a problem by no means confined to the
Lancashire-Cheshire area. For the tripartite concept won wide support among
British geologists until quite recently. Most regions could claim their own
tripartite scheme, even though in most cases it was not possible to demonstrate
that actual superimposition of units at a given locality. The unfortunate
practice was to name boulder clay units, upper or lower, according to their
relationship with the nearest sand unit. At worst such a relationship was
inferred or even guessed at. One of the results of the adoption of tripartite
classifications was the recognition of readvances. Readvances have, however,
been subject to more critical examination since then, and many of them
invalidated. Similar problems arising from simple views of lithostratigraphy
occur elsewhere in Europe and in North America (Chapter 8).

Lithostratphic units of member status are those next in rank below a formation,
and must always be some part of a formation (Table 4-2, Fig. 4-4). A member
must possess lithologic properties which enable its distinction from adjacent
members of the formation (Fig. 4-4). Sometimes, as in the case of the Robein
Silt (formation) in Illinois, formations are not subdivided into members
(Willman and Frye 1970). As mentioned previously the criterion for subdivision
should be that better understanding should always result.

A bed is the smallest formal unit in the hierarchy, although terms such as
laminae may be used when appropriate. A bed should be distinct lithologically
from units above and below it, and the term is customarily applied to strata
a centimetre to a few metres in thickness. Distinctive beds may be called
marker beds, for example, the three Pearlette Ash deposits in the U.S.A.
(Chapter 2).

Every lithostratigraphic unit should have a stratotype. Table 4-1 is an
example of a stratotype of formational status from Illinois (Subcommission on
North American Quaternary Stratigraphy 1977). Description of the Jessup
Formation (Table 4-1) includes statements on the basic requirements of a
stratotype : that is, the history of the concept, locality and region details,
type of stratotype, lithologic characteristics - note the use of colour
description based on the Munsell Standard (Munsell Soil Colour Chart 1954)

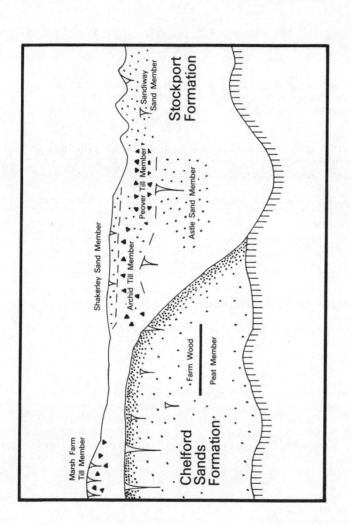

Fig. 4-4. Lithostratigraphic units in the Lancashire-Cheshire Plain (after Worsley 1967). V-symbols show ice-wedge horizons. Stockport Formation is glacial (Irish Sea Ice). Chelford Sands Formation is an alluvial-aeolian complex. Farm Wood Peat represents Chelford Interstadial (Brørup)

TABLE 4-1   Description of the Jessup Formation
(Mainly loam till of Illinoian and so-called
Kansan age in southern Indiana (After The
Subcommission on North American Quaternary
Stratigraphy 1977)

Named and defined by: Wayne, W.J., 1963, p. 52-55,
Pleistocene formations in Indiana: *Indiana Geol.*
Survey Bull. 25.

See also: Wayne, W.J., 1970, p. 79-80, Jessup Formation,
in R.H. Shaver and others, Compendium of rock-unit
stratigraphy in Indiana: *Indiana Geol. Survey Bull.*
43.

Gooding, A.M., 1963, Illinoian and Wisconsin glaciations in
the Whitewater Basin, south-eastern Indiana, and
adjacent area: *Jour. Geol.*, v. 71, p. 665-682.

Gooding, A.M., 1966, The Kansan glaciation in southeastern
Indiana: *Ohio Jour. Sci.* v. 66, p. 426-433.

Bleur, N.K., 1976, Remnant magnetism of Pleistocene sediments
of Indiana: *Proc. Indiana Acad. Sci.*, v. 85, p. 277-
294.

Named from: Town of Jessup in Parke County, Indiana;   20 km
northeast of Terre Haute.

Holostratotype locality: Stream bank along tributary to
Straners Branch, about 10 km northeast of Jessup,
Catlin quadrangle, Lat. 39°40'48"W.

Definition: The Jessup Formation consists mainly of till
with associated silt, sand, and gravel. Two members
are recognized: the Butlerville Till Member (upper)
which includes the greater part of the formation, and
the Cloverdale Till Member (lower) which is recognized
in only a few places. The members are distinguishable
primarily by a palaeosol, the so-called Yarmouth soil,
which is developed in the lower member either in the
till or in associated nonglacial sediments that lie on
top of the till.

Holostratotype section: About 15 m exposed, principally
till but including some thin beds of silt and sand.
Midway in section is a bed of peat 0.5 m thick and C^{14}
dated at >36,000 BP. The peat marks the contact
between two till members: the Butlerville Till Member
(upper) mainly is yellowish-brown (10YR hue): the
Cloverdale Till Member (lower) includes more olive
hues (5Y, 2.5Y).

Boundaries: Base: At the stratotype section the Jessup
     Formation is thought to rest on bedrock of Penn-
     sylvanian age, which crops out nearby. In most
     exposures in which the base of the Jessup Formation
     is seem, the formation rests on bedrock of Palaeozoic
     age.
     Top: The Jessup Formation is at or near the surface
     over a large area; commonly it is overlain by varied
     facies of Atherton Formation, particularly the loess
     facies. In west-central Indiana, the Jessup Formation
     has been observed beneath till of the Trafalgar
     Formation in many places.

Additional information: The internal stratigraphy of Jessup
     Formation remains largely unstudied. Wayne (1970)
     observed that the formation includes several thin beds
     of fossiliferous silt, and made tentative correlations
     with multiple tills described by Gooding (1963, 1966)
     in southeastern Indiana.

Regional extent: The Jessup Formation, principally the
     Butlerville Till Member, is widespread in southern
     Indiana where it underlies an area of about 9,000 km^2.
     Over about half this area it is an extensive sheet-
     like deposit as much as 30 m thick; in the remaining
     area, the till is patchy and generally thinner. It
     is widely present also beneath the Trafalgar Formation
     in central Indiana.

Geologic age: The greater part of the formation, presumably
     the Butlerville Till Member, is contiguous with till
     in Illinois that occurs at the surface and has generally
     been regarded as Illinoian in age. The Cloverdale
     Till Member was assumed by Wayne (1963, p. 56) to be
     so-called Kansan in age. The palaeosol developed at
     the top of the member is not convincing, however, and
     associated sediments have been shown to be mostly of
     normal DRM polarity (Bleuer, 1976). The age of this
     member thus remains in doubt.

which enables ready appreciation of the exact hue instead of being confronted
by a vaguely defined colour, the stratotype boundaries, regional extent, and
data on its geologic age.

How many of the lithologic units which figure prominently in continental
classifications (Chapter 2) have been defined in such terms? The answer is,
unfortunately, very few. This is not altogether surprising because, recently,
in a north-west European context, it could be said that :

> the definition of lithostratigraphic units and their
> separation from chronostratigraphic thinking is some-
> thing new in Quaternary stratigraphy (Luttig *et al*
> 1969).

This would not be true for North America, where the means of subdividion has
for long been a matter of debate (e.g. Morrison 1965). While few opportunities
exist for proposing new formations in pre-Quaternary rocks (Hedberg 1976),
most Pleistocene and Holocene rocks have yet to be formally defined. Two
further kinds of units can be included under the broad heading of lithostratigraphy although they are conceptually separate. These are the *soil
stratigraphic unit*, and the *morphostratigraphic unit*, both encountered in the
discussion on central North America and the Alps respectively in Chapter 2.

The importance of fossil soils (palaeosols, paleosols) for subdividing strata
and from which environmental inferences may be drawn, is self-evident (Table
4-2). Some have argued for their adoption as major time-parallel units of
widespread extent (Morrison 1965), a point returned to later. A soil
stratigraphic unit is defined as :

> a soil with physical features and stratigraphic relations
> that permit its consistent recognition and mapping as a
> stratigraphic unit (American Code 1961).

It differs from a lithostratigraphic unit in that it is developed *in situ*
from underlying lithological units which may be of different ages; moreover
it originated later than them, and under different environmental conditions.
Soil stratigraphic units should be defined according to the same procedure
as for lithostratigraphic units.

Morphostratigraphic units are of *informal* status only. They are not lithostratigraphic units, and their terminology should not be applied with any
stratigraphic implication (Richmond 1959). Willman and Frye (1970) adopt
them in their classification of the Quaternary of Illinois (Table 4-2), and
define a morphostratigraphic unit as 'a body of rock that is identified
primarily from the surface form it displays'. An important aspect of their
definition is that it includes both landform and lithology. Their map of
moraine ridges in north east Illinois (Chapter 2) indicates the enhanced
detail that such units make available.

Morphostratigraphic concepts are also applicable to terrace sequences, but
should never be substituted for any other kind of stratigraphic unit.
Examples of such usage as found in Chapter 8.

Biostratigraphy

The object of biostratigraphy is to group strata into units on the basis of their contained fossils. This may be used on differences in fossils, or differences in the relative abundance of certain taxons*. That is, definition is based on observable features, and as such is objective. Rocks which do not contain fossils cannot be grouped into biostratigraphic units.

Biostratigraphy is not synonomous with lithostratigraphy. In certain cases lithostratigraphic and biostratigraphic units may coincide, but the latter often overlap the former. A biostratigraphic unit is defined as :

> a body of rock strata unified by its fossil content or palaeontological character and thus differentiated from adjacent strata. A biostratigraphic unit is present only within the limits of observed occurrence of the particular biostratigraphic feature on which it is based (Hedberg 1976).

The term biozone may be used for a biostratigraphic unit, though the term zone should not be used without a descriptive prefix. Four types of biozone are mentioned by the Stratigraphic Guide (Hedberg 1976) :

> *Assemblage-Zone* : strata with a distinctive fossil assemblage
>
> *Range-Zone* : strata representing the stratigraphic range of a selected element
>
> *Acme-Zone* : strata based on abundance or development of certain forms
>
> *Interval-Zone* : stratigraphic interval between two biozones

Terms, such as acme (epibole) zone, are not recommended by some authorities for use in the Quaternary because they appear to have no practical use (e.g. Luttig *et al* 1969). Most frequently used in Quaternary work, however, is the assemblage zone (cenozone). It is based on fauna or flora and is ecologically controlled in the first instance in that the assemblage is conditioned by, and reflects a particular environment (Fig. 4-5). It may include fossils entombed together, fossils that lived together, died together, or accumulated together. Assemblages must be scrutinized to identify forms that may have been washed in. Older fossils are often reworked in this way. Less frequently younger forms may be introduced.

An assemblage zone is named after its principal taxon or taxons (Fig. 4-5), and as with other stratigraphic units is best conceived with reference to a stratotype. Table 4-3 illustrate the description of a stratotype for a pollen biozone in Minnesota, and is exemplary in that respect (Cushing 1967).

---

* Following the International Stratigraphic Guide (Hedberg 1976) the plural of taxon is taken as taxons. Taxon is not a classical Greek word but an English invention. The Botanical and Zoological codes give the plural a pseudo-Greek flavour - hence, taxa.

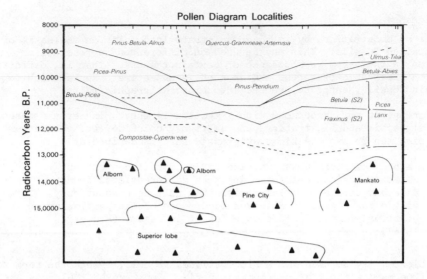

Fig. 4-5.   Correlation of pollen zones and glacial phases in
eastern Minnesota.   Glacial lobes indicated (after Cushing
1967)

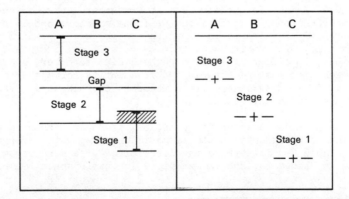

Fig. 4-6.   Unit and Boundary Stratotypes (after
Hedberg 1976)

TABLE 4-3   Description of stratotype for a Pollen
             Assemblage Biozone, *Betula-Picea*, Weber
             Lake, Minnesota, U.S.A. (Cushing, 1967)

Type locality and section: Weber Lake, Lake Co., Minnesota.
Core C:1, zone 2 of Fries (1962).

Description: *Betula* is the most abundant pollen type in
this zone, and *Picea* is the next most frequent. *Salix*
is relatively important (more than 3%). *Larix* and *Abies*
are absent or infrequent, and pollen of other trees and
shrubs (including *Pinus, Juniperus/Thuja, Populus, Quercus,
Ulmus,* and *Alnus*), is relatively unimportant (less than 5%
each). Nonarboreal pollen is moderate (20 to 40%);
Cyperaceae and *Artemisia* are the principal contributors.

Contacts: The lower boundary of the zone is marked by an
abrupt increase of *Betula* pollen from less than 5% and a
corresponding decrease in nonarboreal pollen from more than
50%. The upper boundary is placed at a decrease in *Betula*
to less than 10%, an increase in *Picea* to more than 30%,
and at the beginning of a continuous curve of *Latrix* (more
than 1%).

Other occurrences: The zone is present in two other cores
(S: 1 and S: 2) from Weber Lake (Fries 1962). At Spider
Creek (Baker 1965) it is identified between 875 and 890 cm,
although *Picea* there exceeds *Betula,* as it does also at
Glatsch Lake.

Thickness, age and extent: At Weber Lake the zone is 50 cm
thick; it is dated at 10,500 and 10,200 BP. At Spider
Creek and Glatsch Lake the zone is 15-20 cm thick. The zone
is known only from St. Louis and Lake Counties, Minnesota.

Remarks: Macrofossils of *Betula cf. B. glandulosa* are
reported from this zone at Weber Lake (W.A. Watts, unpub-
lished; cf. Fries 1962) and Spider Creek (Baker 1965).

Due to the climatic control of environments, subject to migration of facies, assemblage zones are time-transgressive. This is illustrated in Figure 4-5, and discussed in the caption to that figure. Some assemblage zones, however, may be of widespread extent, and are time-parallel within broadly specified time-bands : for example some biozones recognized in deep-sea cores (Chapter 6).

Problems of homotaxis are acute with respect to assemblage zones, though on occasion it is possible to indicate general age on the basis of extinction. In Britain, for example, Early Pleistocene assemblages are characterized by Tertiary taxons, those of the Middle Pleistocene are not.

Quaternary biostratigraphic zonation has achieved maximum usefulness in the field of pollen analysis, despite persisting difficulties in comparing past and present floral environments, and is considered in Chapter 6.

Chronostratigraphy

The purpose of chronostratigraphy is to organize strata into chronostratigraphic (time-stratigraphic) units which correspond to intervals of time called geochronologic units (Table 4-4).

A chronostratigraphic unit is defined as :

> a body of rock strata that is unified by being the rocks formed during a specific interval of geologic time. Such a unit represents all rocks formed during a certain span of Earth history and only those rocks formed during that time span. Chronostatigraphic units are bounded by isochronous surfaces (Hedberg 1976).

TABLE 4-4   Chronostratigraphic and Geochronologic Units

Chronostratigraphic Unit	Equivalent Geochronologic Unit
Eonothem	Eon
Erathem	Era
System	Period
Series	Epoch
Stage	Age
Chronozone	Chron

The hierarchy of chronostratigraphic units is shown in Table 4-4. If necessary position within a chronostratigraphic unit can be referred to using the terms lower, middle and upper. The corresponding terms indicating position in a geochronologic unit are, early, middle and late.

The *stage* is the basic working unit of chronostratigraphy. Stages are best defined by boundary stratotypes, as it is only rarely that unit stratotypes are available (Fig. 4-6). Stratotypes should be within sequences of continuous

deposits, and associated with marker boundaries that can be traced as
recognizable features away from the type locality. As many means as possible
should be brought to bear on such correlation. Whereas stages in pre-
Quaternary time last from between 3 to 10 Ma, Quaternary stages are measured
in thousands of years. Especial difficulties are encounted at this high
level of precision. Stage names should normally be derived from a geographic
feature, with the adjectival form being used, ending in 'ian' or 'an': for
example, Hoxnian or Wisconsinan. A substage is a subdivision of a stage and
is defined by boundary stratotypes.

The lowest division is the *chronozone*. This may be based on a lithological
formation, or member, or on a biozone. For example, a biozone based on Pollen
Biozone Hoxnian II will include all rocks, regardless of whether they are
fossiliferous or not, which are the same age as that Biozone. The biostratigraphic unit on which the chronozone is based may correspond to the time
span of the stratotype of that unit, whereupon it will be permanently fixed;
or, it may correspond to the total time span of the unit (often greater than
the stratotype), hence is amenable to modification with new discoveries.
Thus the chronozone may be the type Hoxnian Biozone II, or the chronozone of
the Hoxnian Biozone II. This is a distinction of some importance because a
biozone defined at a type site is inevitably diachronous when traced farther
afield, and the time span between the isochronous surfaces at their extreme
development, earlier, and or later, than at the type site, at whatever
localities, define a larger interval of time. In the case of this example it
cannot be demonstrated whether or not the Biozone at a given locality overlaps
in any way that of type locality. Thus units where independent dating is not
possible should not ideally be subdivided into chronozones, but substages
defined in terms of boundary stratotypes. Chronozones gain maximum usefulness
within the range of radiocarbon dating (Chapter 6).

CHRONOSTRATIGRAPHIC CLASSIFICATION OF THE QUATERNARY

Arising from the earliest subdivisions of glacial deposits in mid-latitude
regions it has become customary to employ climatic changes as the principal
criterion, namely by recognising glacials and interglacials. This seemed
most logical to early workers because till sheets, separated by non-glacial
deposits, could be mapped widely, and shown to lie in a stratigraphic
sequence. Despite developments since, climatic change remains the most
realistics framework for subdivision. Further units of classification were
added : these are, interstadials, representing periods of climatic amelioration during glacials; and stades or stadials, which represent the colder
episodes between the interstadials.

Glacials and interglacials, however, are inferred events. Moreover inference
is not always uncomplicated : sediments are often ambiguous indicators of
origin (Chapters 8 and 9), and even more so of contemporaneous climates.
Regional variability adds further complications : the climatic amelioration
in the middle of the Last Glaciation is considered to represent an interglacial in Siberia, but merely an interstadial in western Europe and Central
North America. Classification, therefore, operates at both local and global
levels. The need for detailed local sequences can hardly be denied, but
these are invariably influenced to some extent by classifications of wider
applicability. At one time the Alpine model was a standard. Now perhaps
the global framework should be the isotopic record determined from ocean
cores. The finely detailed record revealed by these (Chapter 3) places a
new and considerable burden on those classifying continental deposits. In

particular, given the relatively large number of potential stages, the
dangers of homotaxis are greater than hitherto appreciated.

Although inference of former temperatures cannot always be realistically
separated from the effects of former precipitation, it is generally agreed
that the temperature record should be the basis for subdivision. Points
should be defined on the temperature continuum which may be adopted as
boundaries for glacials and interglacials. If possible these should be
defined from continuous sequences of data, though unfortunately, all too
frequently, such data is additive, and overlapping, and combined into a
synthetic curve. Furthermore the principal divisions should correspond to
lithologic entities, and boundaries therein. But how should the boundaries
be placed? And on what evidence should they be recognized?

Several possibilities exist for drawing the boundaries. These are shown on
Figure 4-7 which depicts several sine-curves of temperature change through
time. After such boundaries have been drawn, it is pertinent to ask whether
they represent the limits to chronostratigraphic units or simply units of
climatic change that have been inferred to have existed. This crucial
distinction is discussed below.

The following possibilities for defining glacials and interglacials exist
(Fig. 4-7) :

> (a) On the most theoretical level glacials and interglacials may
> be defined using any combination of the units shown on curve a.
> These have been placed at thermal maxima and minima, as well as the
> mid-points between them.

> (b) Boundaries can be fixed at temperature maxima. A
> stage then consists of a glaciation and the following interglacial.
> It has been suggested, for example, that the Illinoian Glaciation
> can be combined with the Sangamon Interglacial (Morrison 1965). An
> advantage of this is that the boundaries, in this particular case,
> consist of palaeosols, which Morrison argued are time-parallel
> units (soil-stratigraphic units).

> A similar practice in Europe is used for divisions at somewhat
> higher rank. The Upper Pleistocene, for example, consists of the
> Eemian Interglacial and Weichselian Glaciation; and the Middle
> Pleistocene, as defined by Woldstedt (1954) and Luttig (1959),
> includes the Holstein Interglacial and Saale Glaciation.

> A drawback of drawing boundaries at temperature maxima only, however,
> is that the units so defined are larger than would be convenient for
> study purposes and correlation. They would include two climatic
> regimes, each radically different from the other, whereas the prime
> objective of climatic classification is to define discrete temper-
> ature events.

> (c) Some boundaries are drawn at mid-points between maxima and
> minima. This, for example, is the procedure followed by workers
> defining stages from deep-sea cores (Fig. 3-5). To some degree it
> is a purely arbitrary procedure, and it should be noted that the
> oxygen isotope curves are not first order indicators of climatic
> change, but merely of ice-volume on the continents (Chapter 3).
> Correlation on an inter-core basis is fairly uncomplicated on this
> basis, but when the continental and oceanic evidence is compared

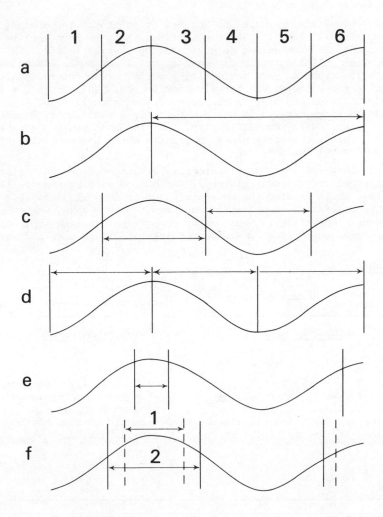

Fig. 4-7.   Various means of defining an interglacial (see text).   The sine
curve represents temperature trend through time

difficulties arise in the 'keying in' of what are climatically
significant horizons. An exception to this would be correlation of
marklines from loess sequences (Kukla 1975) with the terminations of
Broecker and van Donk (1970).

(d) Boundaries may also be defined at points of temperature maxima
and minima. That is, where conditions change : from cooling to
warming, or vice-versa, as advocated by Suggate (1965). The falling
portion of the curve is representative of a glacial, when ice
sheets advanced; and the rising portion, of an interglacial, when
ice sheets decayed and largely disappeared. This procedure is
advantageous because, in the case of the interglacials, the boundary
can be drawn mid-way in what is commonly a continuous sequence of
deposits. With more difficulty, the timing of maximum glaciation
can sometimes be determined, hence the other boundary also fixed in
continental regions.

(e) In north west Europe interglacials are defined on the basis of
inferred vegetational history. Definition is very precise (Table
4-5), as it is also for recognition of an interstadial. For an
interglacial the boundaries are drawn : at the base, where open
tundra vegetation is replaced by forest; and at the top, where the
reverse takes place (Jessen and Milthers 1928. Turner and West 1968).

TABLE 4-5   Definition and sequential development of
            'interglacial', and 'interstadial' (Jessen
            and Milthers 1928)

Interglacial	Interstadial
arctic	arctic
sub-arctic	sub-arctic
boreal	boreal, with summer temper-
temperature, with summer temper-	ature essentially lower
ature at least as high as	than in the Flandrian
during the Flandrian climatic	climatic optimum of the
optimum of the area in quest-	area in question
ion	sub-arctic
boreal	arctic
sub-arctic	
arctic	

Definition is closely tied to a standard of present day conditions,
and can be extended to include other evidence, for example,
palaeosols, or fauna. The base of an interglacial so defined,
corresponds in concept to that adopted for the lower boundary of the
Holocene, and it has been urged that this be taken as a standard
(Luttig *et al* 1969). Also, that the vegetational history and
boundaries established for the Eeemian Interglacial be used as a
standard for other interglacials. Subject to any complications due
to homotaxis, interglacials thus defined are ideally suited on a
local basis. Unlike most continental deposits those on which interglacials are defined are usually continuous sequences where lower
and upper boundaries occur : hence they are unit stratotypes.
Unfortunately the evidence on which boundaries are defined is time-

transgressive, hence when traced from the type region, may become progressively less suitable (Chapter 6).

(f) One of the drawbacks of the interglacial as defined in Europe (above, e) is that, other than in a few cases where its duration may be estimated (Chapter 5), there is no way of relating it to an interval of time on, say, oxygen isotope curves. Does, for example, the Last (Eemian) Interglacial of Europe correspond to all of stage 5 of the deep-sea cores? Shackleton (1969) argued that it did not, and corresponded only to stage 5e. The other peaks within stage 5, being, on this assessment, interstadial rather than interglacial. This is a matter of continuing debate (below).

If the Eemian, as seems likely, only corresponded to stage 5e, it means that it lasted for about 11,000 years, and in oxygen isotope terms only represents the culmination, peak, and subsequent initial decline, of the palaeoglaciation/palaeotemperature curve. Evidence of interglacials from lower latitudes on the other hand indicates a considerably longer time span (e.g. Fairbridge 1972), (Fig. 4-7 : f). This illustrates one of the disadvantages of climatic terms such as interglacial : for such conditions were of unequal duration in different parts of the world.

The requirements of Stage units, that they be drawn if possible in sequences of continuous rocks, and that their boundaries be time-parallel, are not readily met by most Quaternary lithostratigraphic or biostratigraphic units. This was clearly recognized by Flint (1957) discussing the stratigraphy of Central North America :

> The seven physical units (four drift sheets and three soil zones) are grouped into seven stages. Being a time-stratigraphic (chronostratigraphic) unit, a stage implies time correlation; yet obviously the till sheets, at least, are time transgressive. The term *stage* is applied to all seven units with the realization that upper and lower boundaries are not adequately fixed (Flint 1957).

In discussion ranging beyond North America, Morrison went even further :

> I know of no single type section in North America or Europe that suffices to express the full time range of any of the standard Quaternary 'glacial' stages or substages (Morrison 1965).

Herein lies a major dilemma : whereas global units, such as those established by deep-sea records are based on, and involve continuous sequences which span the entire climatic events as defined, continental units inevitably, except in some exceptional cases, lie inside the large entities, and moreover have time-transgressive boundaries within these.

That is not to say that local classification cannot proceed perfectly adequately, as witness numerous instances, notably through using boundary stratotypes, which at a given locality are time-parallel. But instead of recognising such a situation for what it is, and proceeding pragmatically within its constraints, a new unit was proposed and given formal status - the geologic-climate unit :

> A geologic-climate unit is an inferred widespread climatic
> episode defined from a subdivision of Quaternary rocks
> (American Code 1961).

It has several synonmym,some of which long antedate its formalization as a
conceptual stratigraphic unit : climatic units (Frye and Richmond 1958),
climate-stratigraphic units (Richmond 1959), 'units based on climatic change'
(West 1968), glacial-stratigraphic units (Flint 1971), climatostratigraphical
units (Mangerud *et al* 1974). Earlier Flint (1957) had been less definitive,
merely commenting that schemes used in Europe and North America are based on
lithologic units consisting of alternating glacial and nonglacial deposits.

Boundaries of the geologic-climate unit (American Code 1961) were meant to be
those of the stratigraphic unit on which it was based. It was realized that
these would be time-transgressive. Further definition included :

> A glaciation was a climatic episode during which extensive
> glaciers developed, attained a maximum extent, and receded.
>
> An interglaciation was an episode during which the climate
> was incompatible with the wide extent of glaciers that
> characterized a glaciation. A stade was a climatic
> episode within a glaciation during which a secondary
> advance of glaciers took place. An interstade was a
> climatic episode within a glaciation during which a
> secondary recession or stillstand of glaciers took place
> (American Code 1961).

These terms were, of course, of long standing currency, but their re-emphasis,
in formal terms, was somewhat restrictive geographically, because as was
pointed out, most of the world was unglaciated during the Quaternary (Suggate
1962).

More often than not, the terms defined, glacials, interglacials, stades
(stadials) and interstadials were accorded stage status by many workers.
However, most of the time, they fail lamentably to meet the requirements of a
stage as recommended in any stratigraphic guide or code. Are they stages, or
glacial and interglacial ages? If ages, how is it possible for them to be so
without proper chronostratigraphic definition? No doubt some of these
difficulties are more apparent than real, and are to some extent based on
semantic confusion. But this discussion serves to identify what is a
fundamental lack of agreement in defining these everyday concepts. Usually
expediency prevails, and while tacit recognition is paid to difficulties and
inconsistencies (see Flint quotation above), stage status is accorded these
units. Thus we have the 'Nebraskan Glacial Stage', and 'Sangamon Interglacial
Stage' (Flint 1971). Neither has an adequate chronostratigraphic definition.

It is not surprising therefore that the potential for indiscriminate mixing
of lithostratigraphic, biostratigraphic, chronostratigraphic and geologic-
climatic units, was, and unfortunately, still is rife.

Usage in Central North America and the British Isles

Chamberlain's and Leverett's two fold classification of mid-continental deposits,
into glacial and interglacial, lasted for nigh on forty years before they were
formalized (American Code 1933). Their interglacials represented times when

weathering profiles (palaeosols) formed, while glacials occurred in between
such times (Fig. 4-7, caption). There can be no doubt that according to
their conception, and those of subsequent workers, that their terminology
was meant to represent chronostratigraphic units at stage rank (e.g. Willman
and Frye 1970).

Unfortunately, whatever the pragmatic merits and demerits of their scheme, a
certain amount of confusion ensued when they became partly used as lithostratigraphic, chronostratigraphic and what were to become later geologic-
climate units. Thus :

> By long-continued custom in the United States the time
> covered by a Pleistocene subdivision of formational rank
> is called a *stage*, and the time covered by a Pleistocene
> subdivision of member rank is called a *substage* (American
> Code 1933).

For a small area such a recommendation worked. But the terminology was
applied on a continental scale. Thus lithostratigraphic units, with lower
and upper boundaries that were clearly time-transgressive, were accorded
chronostratigraphic status. This confusion by no means only applies to the
continental United States : it is part and parcel of investigations in all
the formerly glaciated regions, particularly when 'count from the top'
procedures are applied : for example the Irish Sea region (Bowen 1973. 1977).

The unsatisfactory position arising from the 1933 definitions was recognised by
the 1961 American Code which attempted to remedy the situation by introducing
the geologic-climate unit. If nothing else, this at least clarified the
issues so that of the subdivisions the question could be posed : are they
stages or geologic-climate units? Many continue to use them as valid stages,
but in a strict stratigraphic sense they cannot be so. Full and adequate
definition, or redefinition, as chronostratigraphic units is required at
stratotypes.

An attempt to meet some of the difficulties inherent in classification has
been made in the British Isles (Mitchell *et al* 1973). Units of stage rank
are defined at stratotypes by drawing boundaries at their base only : that is,
using boundary stratotypes for lower boundaries, although this is not
explicitly stated. Unfortunately, even though uniform criteria could have
been applied throughout, two criteria for boundaries are used. Interglacial
stages are defined on biostratigraphic (pollen)evidence, while glacial (cold)
stages are defined on lithologic criteria - for example the base of a till.
It means almost by definition that major gaps exist, despite the fact that the
sequence is shown as a continuum on tables and climatic curves (Chapter 2).

Thus the base of the Hoxnian is that of the lower boundary of Pollen Assemblage
Biozone HI, the base of the succeeding Wolstonian Stage - the lower boundary
of the Baginton-Lillington Gravels, and the base of the succeeding Ipswichian
Stage - lower boundary of Biozone Il. In this way it is quite possible for
major gaps to occur between the top of the Hoxnian (not defined, though it
could be) and the base of the Wolstonian lithologic unit; and also between
the top of the topmost Wolstonian member, the Dunsmore Gravel, and the base
of the Ipswichian.

A far better means of definition could have defined both the base and top of
the interglacials; the Wolstonian could then refer to the time interval

between the end of the Hoxnian and the commencement of the Ipswichian. In
this way gaps could be eliminated.

THE LAST INTERGLACIAL, LAST GLACIATION, AND THE HOLOCENE

Certain concepts are so commonplace that they have gained widespread acceptance despite the fact that they lack adequate definition. Among the most
frequently used are the three discussed here.

The Holocene

Everyone is familiar with the concept of postglacial time, the interglacial
of the present day. But how should its lower boundary be defined? In a
review of the many criteria urged, and adopted in some areas, Morrison (1969)
showed how estimates of its age ranged from 20,000 to 4,000 years ago. Some
of these are, on theoretical grounds, eminently reasonable (cf Fig. 4-7).

At one time workers in the Mississippi delta region believed the boundary
should correspond to the time of glacio-eustatic sea level low at the maximum
of the Wisconsin glaciation : that is, about 20 ka BP. Others sought a
stratigraphic marker of widespread extent, such as a palaeosol. The base or
top of the Allerød could fulfil such a role. It was also drawn at the top of
the Valders till, about 11.5 to 11 ka BP in the Lake Michigan Lobe, or
alternately, at the contact of the Cochrane Till and James Bay lowland deposits
in Canada, dated to about 7 ka BP. Other boundaries that could have been taken
(cf Fig. 4-7) include the peak of the postglacial marine transgression about
5.5 to 5 ka BP, or the thermal optimum ('altithermal') Atlantic warm interval
of Europe about 8 to 5 ka BP.

In the event, the Holocene Commission for INQUA decreed that the boundary
should be placed at 10,000 radiocarbon years ago, at a temporary level approximately half way between the cold of the Last Glaciation and the postglacial
thermal optimum. Undoubtedly this reflects a tendency to accepting the
pollen analytical standard of Europe as one of wide significance, in that it
may be inferred as a first order indication of climate.

By defining the base of the Holocene in this way the Subcommission have
anticipated the discovery of a suitable stratotype. Can such a unit of
geochronologic time exist without a chronostratigraphic standard of comparison?
Clearly it cannot. Grounds of expediency have won the day and the extra-
ordinary situation exists whereby the search for a stratotype continues. That
proposed by Morner (1976) has been rejected, but may, nonetheless be studied
as an example of how such a detailed proposal should be made. Fairbridge
(1976) surveys those areas likely to yield such a stratotype and some of the
difficulties involved in some environments. Localities along the Swedish
coast are presently being evaluated.

The dilemma of a defined stage (age) without its stratotype lends support to
those who maintain that such stratigraphic procedure is unnecessary and on
occasion a hindrance. Yet, when all is said and done, without a standard,
what is there as a yardstick for comparison? In North America the crucial
date seems to lie about 13 ka BP in northern Canada about 6 ka BP, and in low
latitudes between 11 and 9 ka BP. In this light the compromise date of
10,000 years ago does not seem unrealistic.

The boundary between the Last Interglacial and Last Glaciation

Two schools of thought obtain on the length of the Last Glaciation. One terminates the Last Interglacial at 116 ka ago, the other at 70 ka ago. Thus according to one estimate it lasted for 106,000 years, but only 60,000 according to the other.

Crucial to this argument is the status of the temperature maxima and minima which may be seen on Emiliani's (1961) oxygen isotope curve pertaining to stage 5 (Fig. 4-8), also revealed on other workers analyses (e.g. Shackleton and Opdyke 1973). It was formerly customary to include all of stage 5 within the Last Interglacial, with its upper boundary dated as ca. 70,000 years ago. This date appears to have been fixed because some of the isotopically enriched radiocarbon dates for interstadial episodes in the Last Glaciation gave estimates of, for example, 65 to 68 ka BP, for the Amersfoort Interstadial (Fig. 2-10) (see also Chapter 5).

Subsequent to this, however, it was argued that the Eemian Interglacial of Europe, as characterised by Jessen and Milthers' (1928) (Table 4-5) definition, only corresponded to isotope stage 5e (Shackleton 1969). It was ended by a fall in the oxygen isotope curve corresponding to a sharp drop in palaeo-temperature, incapable of recognition in the pollen record until the end of the Eemian. A corollary to this is that the subsequent peaks in stage 5, namely 5c and 5a, represent interstadials, possibly Amersfoort and Brørup, notwithstanding the radiocarbon determinations on those episodes (Chapter 5). Further evidence has clarified the nature of these episodes.

Using transfer function equations on fossil populations (Chapter 6) it was shown that subtropical surface waters only reached as far north as ca. latitude 52 degrees north on one occasion during stage 5, shortly after 125 ka ago : that is, corresponding to stage 5e. But transitional surface waters swept northwards on two subsequent occasions. Between these three episodes of comparative warmth, subpolar water shifted southwards in the north Atlantic (McIntyre, Ruddiman and Jantzen 1972). At such times of cooling, 110,000 and 92,000 years ago, ice rafted debris was recorded in deep sea cores. At 110 ka the cooling represents two-thirds of a swing towards full glacial conditions (Sancetta, Imbrie, Kipp, McIntyre and Ruddiman 1972). In the Gulf of Mexico warm peaks at 120 and 100,000 years were followed by abrupt cooling at 90,000 years (Kennet and Huddlestun 1972). 116,000 years ago sea level fell by 71 ± 11 m in Barbados (Mathews 1972).

In continental regions the period is recorded by three palaeosols, of inter-glacial rank, separated by glacial advances as extensive as that of the Last Glaciation, in the Rocky Mountains (Richmond 1972). In central Europe loess accumulated soon after 115,000 years ago (Kukla 1975).

The data adduced seem overwhelming in support for regarding only stage 5e as representative of the Last Interglacial. Yet, if all of stage 5 is not adopted as interglacial, with included climatic oscillations, how are earlier inter-glacials in the oxygen isotope record to be interpreted? It may be that due to proximity to the present, with attendant richness of evidence, that the Last Interglacial must be accorded slightly different interpretation than for earlier ones, at least for the present. For the fact is that of the stages shown on V28-238 (Fig. 3-4) stage 19, 13 and 7, do not reach the isotopic values of the later peaks of stage 5, while stage 15 barely attained those levels. It has been argued that for purposes of global correlations the Last

Interglacial is best defined so as to include all of stage 5 (Suggate 1974).

For present usage, however, it would seem that a consensus view would adopt only stage 5e as the Last Interglacial. The Last Glaciation thus commences 116,000 years ago (Kukla, Mathews and Mitchell 1972).

Classification procedure in the Quaternary then is by no means uncomplicated. Much of the data does not easily lend itself to the requirements of properly defined stratigraphic units, certainly at chronostratographical level. Bound- aries are often drawn on grounds of expediency consistent with the available evidence. This pragmatic approach is often the only available means, hence is provisional, pending new data. These situations should be recognised as such and improper usage of stratigraphic terminology avoided.

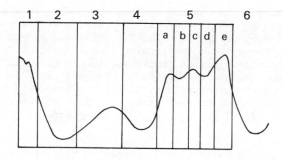

Fig. 4-8. Oxygen isotope curve for stages 1 to 6 (Emiliani 1955)

Chapter 5
GEOCHRONOMETRIC DATING

Stratigraphic investigation allows dating by *relative* means : i.e. 'older than', or 'younger than'. Geochronometric dating, however, is an artificial periodical scale (for it cannot be discovered) of equal units of duration. The unit, since radiometric dating commenced, is the year. This can be defined in seconds, based either on Emphemeris (astronomical) Time, or on the Caessium Atom. Zero, or datum year, is taken as 1950, following the lead given by radiocarbon dating.

Geochronometric is preferred to Geochronologic(al) because the latter has been, and is, used to refer to both chronostratigraphic time spans (Chapter 4) as well as radiometrically determined ones. The term *absolute* dating is unsuitable : it implies a degree of achievement hardly consistent with the realities of the majority of dating methods, which, in terms of their present

TABLE 5-1   Geochronometric Dating Methods Applicable to
the Quaternary

	Isotope Half-life*	Method	Range*	Materials
Radiometric	^{14}C 5.7	Decay	50	Organics
			25	Organic carbonate
	^{230}Th 75	Growth	0-200	Organic carbonate
	^{234}U 250	Decay	50-100	Coral
	^{4}He	Growth	No limit	Molluscs, coral
	^{40}Ar	Growth	No limit	Volcanics
	^{39}Ar	Growth	No limit	Volcanics
Floating scale	Amino·Acid Diagensis	Racemization	200	bone, shell
	Glacial varves	annual deposits	4	glacial lake clays
	Organic varves	annual deposits	50	lake clays
	Dendrochronology	annual tree rings	ca. 9	trees
	Palaeomagnetism	polarity reversals	No limit	Volcanics Sediments

*Half-lives and range in 10^3 years

For means of dating deep-sea red clay and Globigerina oozes using ^{231}Pa and ^{230}Th see Broecker 1965.

109

status may be likened to the top of a gigantic experimental ice-berg.

Fundamentally two methods are used (Table 5-1) : (1) that which provides a
measure of the age of a sample, such as K-Ar dating, and (2) that which
provides a 'floating' chronology, and which may be tagged by radiometric
dates, for example, tree ring chronologies or amino-acid diagenesis. The
former (1) depends on natural rates of decay of radioactive isotopes, the
latter (2) mostly on natural processes with annual rhythms.

RADIOCARBON DATING

W.F. Libby was awarded the Nobel Prize for Chemistry in 1960 as a result of
his pioneer work on radiocarbon (carbon-14) dating. The fundamentals of the
method are as follows. Carbon-14 is produced in the earth's upper atmosphere,
is then oxidised and mixed through the atmosphere, where it is taken up
directly by green plant photosynthesis, or less directly by other organisms
through food chains. The amount taken up in this way is in much the same
proportion to stable carbon as exists in the atmosphere. This exchange
ceases on death, whereupon the unstable carbon-14 is subject to radioactive
decay. By measuring the amount of radioactivity in a fossil sample, comparing
it with a modern standard, and armed with a knowledge of its rate of decay
(half-life), it will be possible to calculate the time that has elapsed since
death. This allows the dating of geological events with which the sample was
contemporaneous. It is assumed that the rate of carbon-14 production over
time was constant, which it was not, but this and other difficulties are
discussed later.

Materials for radiocarbon dating include : wood, charcoal, peat, organic mud
(*gyttja*), and also calcium carbonate as in molluscs, foraminifera, and bones.
It is also possible to date certain precipitates such as tufa. The reliability
of such dates varies according to the nature of the sample, its age, and the
environment from which it was collected.

Production and Decay of Radiocarbon

Carbon-14 is constantly produced in the earth's upper atmosphere, where cosmic
ray bombardment generates neutrons. These are uncharged particles and they
cause transmutation in the nucleus of any atom with which they collide. In
this way nitrogen-14 atoms are transmuted into carbon-14 atoms. Carbon-14 is
then oxidised to carbon dioxide in the atmosphere (which amounts to only
0.033%), most of it consists of stable carbon-12 (98.89%), with a smaller
amount of stable carbon-13 (1.11%), and a minute quantity of carbon-14 (1 part
in 10^{12}). The radiocarbon dating method thus deals in the measurement of
very small quantities of that radioisotope; the older the sample, the smaller
the remaining activity, and hence the more difficult the operation.

Variation in the production of carbon-14 occurs both geographically and
geologically. The latter is considered later. Geographical variation occurs
both latitudinally and altitudinally. Cosmic ray intensity is at a maximum
near the poles, and a minimum at the geomagnetic equator. This is because
cosmic rays are deflected by the earth's magnetic field, unless they happen
to be travelling parallel to the lines of force. At the outer limit of the
stratosphere, neutron intensity is zero, but increases to a maximum value
at 15 km, before falling to 3% of the maximim at about 3 km between
latitudes 50 and 90. At sea level it falls to 0.3% of the maximum value.
These latitudinal and altitudinal variations, however, are comparatively
unimportant because the rate of mixing in the atmosphere is rapid. More

TABLE 5-2   Production and decay of Carbon-14 (after
                Burleigh 1972)

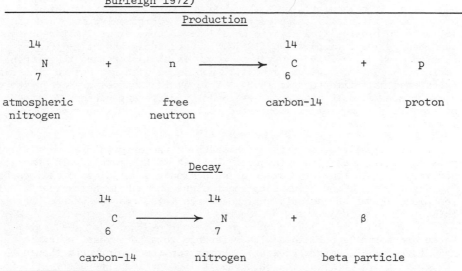

Total mass of atoms shown by superscript figures.
Number of protons in the nucleus shown by subscript figures.

crucial is the variation over time.

Atmospheric carbon dioxide enters the oceans as well as the biosphere, and the carbon exchange reservoir includes atmosphere, surface ocean, deep ocean, biosphere and the humus on the earth's surface. Within each of these the mixing time is rapid. But rates of exchange between atmosphere and ocean vary, and are determined by circulation patterns which may vary over time. For example, there is a measurable defficiency of carbon-14 in the deep ocean, and some deep water samples can give radiocarbon dates of up to 2,000 years. Even surface waters commonly give ages of about 400 years or so. The carbon exchange reservoir is thus a complicated natural mechanism; some of its principal features are considered by Aitken (1974).

All carbon dioxide taken up by green plants during photosynthesis contains a minute quantity of carbon-14. Carbonates such as calcite and aragonite also take up the isotope. At death the uptake of further carbon dioxide ceases, whereupon radiocarbon is subject to radioactive decay as it breaks down into stable nitrogen by a mode of disintegration called beta transformation (Table 5-2). The original nitrogen structure is restored when an energetic beta particle (energetic electron) is emitted. These beta particles are very weak, with a maximum energy of only 160 Ke V, and they are emitted when the radio-carbon atoms decay.

Constant production of radiocarbon in the earth's upper atmosphere is offset

by constant decay. Thus, as Libby theorized, an equilibrium obtains, where
production is balanced by decay. The rate at which carbon-14 decays, its
half-life, is calaculated from the observed rate of disintegration of a known
amount of artificially produced radiocarbon. Originally Libby and his
collaborators calculated this to be 5570 + 30 (the error limit representing
one standard deviation) (Libby 1955). Subsequently the half-life was
recalculated to be 5730 + 40 years, which is 3% larger than the original.
Radiocarbon dates, however, are still published with respect to the originally
calculated half-life (Godwin 1962). This avoids unnecessary confusion which
would ensue if both standards were used by different workers. If necessary
older dates can be readily converted to the new half life, simply by multiply-
ing by 1.03.

After ten half-lives, only one thousandth of the original radioactivity remains;
this is usually taken as a theoretical limit to the method. In practice,
however, because the measurement of such minute quantities of radiocarbon is
impossible, the limit amounts to about eight and a half half-lives. Using a
technique known as isotopic enrichment the method may be pushed back beyond
this.

Dates are normally published with reference to a zero year, which is taken as
1950 (B.P. or BP, Before Present). Dates for the Pleistocene should be cited
in terms of BP; but for the Holocene it is sometimes necessary principally
among archaeologists and prehistorians, to quote dates in terms of B.C. and
A.D. An editorial in *Antiquity* (1972) suggested a further convention. This
arose from the belief that realistic conversion from radiocarbon years to
calendar years was possible (below). Hence it was suggested that a.d., b.c.
and b.p. be used for uncorrected radiocarbon dates, using the old half-life
standard, but that A.D., B.C. and B.P. be used only for corrected calendar
age equivalents. As stated above, this is based on a belief that realistic
conversion is possible : it is not, and this convention is unnecessary and
leads to confusion. The various conversion attempts are discussed below :
suffice it to state at this juncture that the time has not yet come, however
close it might be, for full confidence in such methods.

Measurement

The amount of carbon-14 in a sample is determined by measuring its beta-
particle activity. This is achieved by various counter systems operated over
long periods so as to accumulate a statistical reliability for the number of
counts per sample. It is necessary for a long count because radioactive decay
is a spontaneous process. At least 10,000 counts will be required if an
accuracy to + 1% is desired, which takes about 24 hours.

Although the decay process is spontaneous, the observed rate varies randomly
about a mean value. The probability that an individual carbon-14 atom will
disintegrate within oneminute is 1 in 4×10^9, but with a gram of radioactive
carbon, containing 6.6×10^{10} atoms, the combined probabilities give a more
or less steady rate of disintegration. Radiocarbon dates are, therefore,
always reported in the form of a + error in years. This is sometimes
misrepresented as a time-span into which the true date falls. In fact it
merely gives a range of years within which the true date lies according to a
certain probability. The + error refers to one standard deviation, derived
from the counting errors of the sample, background radiation, and the present
day standard sample. These, of course, are all the result of the random
nature of the decay process. The radioactivity determinations scatter around
a mean value as a normal distribution. Thus there will be a 68% probability

that the true value occurs within one standard deviation; 95% probability
that it lies within two standard deviations; and a 99.7% probability that it
occurs within three standard deviations.

Originally, Libby and his collaborators (Libby 1955) used solid carbon in
their measurements. Samples were converted into carbon dioxide by combustion,
then reduced to black carbon by heating with magnesium, before being painted
to the inner wall of a modified Geiger counter. Subsequently two major
techniques replaced this method, those of proportional gas counting, and
liquid scintillation.

Proportional gas counting uses gases derived from the organic sample. These
may consist of : carbon dioxide (CO_2), carbon monoxide (CO), acetylene (C_2H_2),
ethane (C_2H_6), or methane (CH_4), although carbon dioxide, methane and acetylene
are mostly used nowadays.

The counters consist of a cylindrical metal tube closed at both ends, and
through which passes a fine wire, insulated from the outer wall. Then when a
radioactive disintegration occurs in the gas, a minute voltage pulse on the
wire is detected and recorded automatically. These pulses are proportional to
the amount of energy released in the gas by radioactive disintegrations :
hence the name of the method - proportional gas counting. Proportional
counters, however, cannot be used in the open because of the high background
(environmental) count. They need to be shielded from this radiation, and the
practice is to use several tons of old (pre-atomic age) lead, or other non-
radioactive metal. Further, by surrounding the counter with mercury and an
array of smaller counters, the background count can be reduced to as little as
1 or 2 counts per minute. This compares with a count rate for modern carbon
of 10 to 20 counts per minute. The background count must be subtracted from
overall count obtained, hence it is necessary for it to be monitored carefully
so as to ensure a constant level. Much, however, depends on the sophistication
of the machine :using carbon dioxide at 2 atmospheres pressure, a good machine
can measure radioactivity of a sample aged 40,000 years; an exceptionally
good one may go back to 50,000 years. By using methane, used up to 4
atmospheres, some laboratories can date samples up to 55,000 years (Shotton
1967a).

Liquid scintillation counting uses an organic liquid, such as benzene mixed
with an organic solvent such as toluene. The sample is kept in a closed vial
(20 millilitres or less) and is viewed by two photomultipliers capable of
detecting flashed of light not visible as part of the spectrum. These
'scintillations' are caused by beta-particles from the carbon-14 in the
solvent.

This method has certain advantages over proportional gas counting. The amount
of carbon used is contained in a much smaller volume, so that shielding
against background count is much easier. It is also easier to measure the
radioactivity of the sample, background and that of the standard. Moreover
liquid samples are easier to handle for storage purposes. Liquid scintillation counters were designed to measure large numbers of highly radioactive
samples, but are now suited for smaller numbers of low radioactivity (Polach
1969).

Other methods of measurement have been used, though for various reasons are
not widely adopted. Included are (1) the nuclear-track emulsion technique,

which detects beta particles from the tracks they make in photographic
emulsions; (2) thermoluminescent dosimetry, though problems of background
radiation here seem intractable at present; and (3) by using a mass
spectrometer (Rafter 1974).

Two factors make it impossible to obtain a suitable standard for modern
radiocarbon from present day plants. Combustion of fossil fuels has diluted
natural atmospheric radiocarbon. But operating in another direction, and
more than over-compensating for this effect, has been the release of large
quantities of radioactive material to the atmosphere through the testing of
nuclear weapons.

It is, therefore, necessary to use a standard which is independent of these
effects. That employed is the value obtained by measuring 19th century wood,
making due allowance for its radioactive decay in the meantime. The international reference standard to which this relates is achieved by a biosyn-
thetic oxalic acid (bulk stock held by the National Bureau of Standards,
Washington, D.C.). It so happens that 95% of the 1958 activity of this,
after allowance for isotopic fractionation, is the same as the present value
for wood grown in 1890 (Broecker and Olsson 1959). Thus, by using a modern
synthetic standard, laboratories are able to standardize their results.

Problems of Radiocarbon Dating

There are two principal problems : (1) uncertainties that arise because the
initial carbon-14/carbon-12 ratio of a sample is unknown; and (2) contamination by extraneous carbon when the sample has not remained a closed system
since its death - this is a primary source of error for old samples, and is
one which increases with increasing age.

Uncertainties in the initial $^{14}C/^{12}C$ ratio

This potential source of error is common to all radiocarbon dates. During the
Holocene temporal variation in the production of radiocarbon has been clearly
demonstrated by relating tree-ring chronologies to radiocarbon dates. To
some extent it is possible to correct for these fluctuations. But for the
Pleistocene correction factors can only be approximate estimates in most
cases, although in some a high order of probability occurs.

Plants and other animals in plant-based food chains which obtain carbon
directly from the atmosphere are not subject to any other uncertanties. Yet
it is necessary to appreciate that even green plants do not take up carbon-
14 in the same proportion to carbon-12 as exists in the atmosphere. When
carbon dioxide is assimilated during photosynthesis isotopic fractionation
occurs, so that the actual content of carbon-14 in plants is ca. 3 to 4%
below that of the atmosphere : it amounts to an apparent age different of
240 to 320 years. It occurs because during the photosynthetic process
carbon-12 is more readily assimilated than is carbon-14, hence relative
enrichment of the former obtains. A variation also occurs between different
species (Olsson and Osadebe 1974).

It is, however, possible for this by measuring the carbon-13 content of the
sample, using a mass spectrometer. This stable isotope is present to about
1%, and the expected fractionation of carbon-14 is about twice that for
carbon-13. But since actual variations for carbon-13 range from between -35
to +10 parts per mil (relative to PDB) corrections will sometimes need to be

fairly substantial (Olsson 1974). The values are normalized to carbon-13 =
-2 per mil (PDB), which is the normal value for wood. Several actual examples
are discussed by Olsson and Osadebe (1974) who conclude that such corrections
can only be approximate, for too little is understood about the fractionation
process which would allow exact carbon-13 data to be used for different
materials.

Dendrochronology and radiocarbon calibration

Using tree ring data from American and European trees de Vries (1958) was the
first to show that there had been fluctuations in the production of atmospheric
carbon-14. In subsequent studies a chronology based on tree rings of the
bristlecone pine, (*Pinus aristata*), which grows at 3000 m in the White
Mountains of California, figures prominently. Individual trees can be as much
as 4000 years old, and on death tend to persist in a well preserved state. At
the University of Arizona Tree Ring Research Laboratory, a chronology for
8200 years has been established, with prospects for an extension back to 9000
years (Ferguson 1972).

Bristlecone pines were analysed by three university laboratories : La Jolla
(Californa), Arizona and Pennsylvania. This results at La Jolla allowed
Suess (1965) to fit the scatter of data points with a freehand drawn curve
describing a series of 'wiggles' (Fig. 5-1). The validity of these was
subsequently questioned (e.g. Wendland and Donley 1971), for if correct, it
has serious implication for the dating of archaeological samples, because an
old sample may give a younger date than a more recent one, although this could
be obscured to some extent by the standard deviation involved in the count.
In general the divergence between tree ring dates and radiocarbon dates is not
serious after ca. 3500 BP, but before that time the difference becomes
progressively larger, amounting to as much as 700 years by 4500 BP. The cause
of the fluctuations is not certainly known, but it seems likely that it
relates to changes in the intensity of the earth's magnetic field.

Several calibration attempts have been made, and are summarised by Clark (1975).
Among the more recent ones is that of Damon and his colleagues from 549 dated
samples (Damon *et al* 1974); and also that of McKerrell's, who compared dates
from bristlecone pine chronology with historical data from Egypt (Fig. 5-1).
Because they showed marked discrepancies he argued that systematic differences
exist, and that the bristlecone pine chronology is inapplicable in Europe
(McKerrell 1975). At the present time there is no generally agreed calibration,
although an overall trend is clear. Statisticians have produced smoothed
versions of calibration data : for example, that cited above by Damon *et al* is
one that fits a third order orthogonal polynomial to the data.

Using a 1200 year section of a 2990 year floating chronology (i.e. not tied to
a given calendar datum) from oak trees a team from Belfast University investigated sources of potential error in the radiocarbon dating method (Pearson
et al 1977). Bulk sampling and large annual growth rings allowed the product-
tion of 20 year units of wood, each weighing 180-200 g. This compares with
10 year units weighing about 20 g used with bristlecone pine. The counting
method was that of liquid scintillation, and the following sources of error
were investigated and correction factors established : (1) barometric pressure
changes, failure to correct for which can lead to errors of up to 100 years,
(2) background variations with weight loss, (3) background variations with
purity, (4) efficiency variation with weight loss, (5) differential loss of
sample from mixture and correction to normalised weight, including weighing

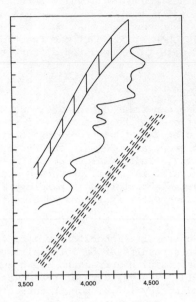

Fig. 5-1.   Relation between
carbon-14 (years shown BP) and
tree-ring years (100 floating
year scale). Top : historical
dates (McKerrell 1975). Middle :
bristlecone pine (Suess 1970).
Bottom : oak trees, north of
Ireland (from Pearson *et al*
1977)

Fig. 5-2.   Two sets of
carbon-14 dates from
Jutland, illustrating a
'hard-water' error (after
Shotton 1972)

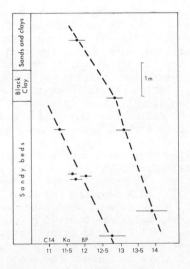

errors, for example, large errors could derive from humidity and temperature effects of plastic caps, (6) efficiency variation with purity, (7) mass spectrometry correction, (8) individual vial difference for efficiency, and (9) individual vial difference for background. This long term study produced the results show on Figure 5-1. It was concluded that significant errors existed in convential radiocarbon measurement techniques. But that they could be overcome so that the overall one standard deviation precision could be reduced to less than 25 years. By increasing the length of the count, it could be reduced to 12 years in 10000. Their regression line (Fig. 5-1) lies within 1% of that produced by Damon *et al* (1974), but does not exhibit the wiggles predicted by Suess.

Problems relating to aquatic plants and carbonates

Aquatic plants and carbonates, such as molluscs, give radiocarbon dates that reflect the particular reservoir whence they obtained their carbon. As such they can only be apparent ages and require investigation before a correction can be applied. This is not only true for continental environments, where the principal source of variation is river and ground water carbon, but also for the marine environment. In rivers the $^{14}C/^{12}C$ ratio can be as much as 40% lower than that of the atmosphere due to 'old' carbon from limestone and other calcareous rocks. In the oceans geographical variations in the ratio occur. Isotopic fractionation also takes place between carbon dioxide and ocean carbonate. An enrichment of carbon-14 of 5% occurs, which goes some way towards compensating for the difficency attributed to variable mixing rates in the oceans. Radiocarbon assays on shells, which were collected before nuclear weapon testing starting, gave apparent ages of 440 years for the Norwegian coast, 510 years from Spitsbergen, and 750 years from Ellesmere Island, Canada. It was shown that if isotopic fractionation is corrected for by normalising according to $^{13}C = -25^{\circ}/oo$, then the entire apparent age should be subtracted to obtain the true one. But if normalized to $^{13}C = 0^{\circ}/oo$ (the average for marine shells) , then a correction of ca. 410 years should be subtracted (Mangerud and Gulliksen 1975).

The carbon-14/carbon-12 ratio will vary in a fairly predictable way in the oceans, but is far less predictable in terrestrial aquatic systems because it is derived from a variety of sources : atmosphere, humus, and dissolved limestone and other calcareous rocks. It is also affected by the nature and amount of vegetation cover, amount of carbonate in the soil, rainfall and temperature. Unless a correction factor is applied, all such dates must refer to maximum ages only. Such correction should take into account : a measurement of the ratio in a representative water sample if possible, and, or, an estimate of how it may have varied over time. In other words detailed geochemical and hydrological case histories are necessary before a date can be treated as rather more than an apparent age. In some cases it is possible to estimate the effect of the factor on the radiocarbon date, but in others the situation is more complex. The following examples are presented to illustrate some of the problems involved. It is perhaps salutory to realize that such uncertainties are not even considered by some users of the dates, who accept them at face value.

In north-east Finland carbon-14 dates did not always concur with expected age on the basis of pollen analysis. Dates from one locality, Säynäjälampi, were invariably 1000 to 3000 years too old when compared with pollen dates zones from other sites (Table 5-3). Measurement of the carbon-14 activity of plants and muds from the lake site compared with the activity of adjacent birch trees, showed that the lake waters were enriched in calcium content relative to local rivers and lakes. It was concluded that a hard water effect accounted for the anomalous dates (Donner, Jungner and Vasari 1971).

TABLE 5-3   Pollen Zones and Radiocarbon Dates from North-
East Finland (after Donner *et al* 1971)

Pollen Zone	Sites and Radiocarbon dates			
	Maaselkä	Meskusjärvi	Maanselänsuo	Säynäjälampi
Hl_4a			$1,900 \pm 130$	$3,450 \pm 170$
Hl_3b			$6,500 \pm 180$	
Hl_3a				
Hl_2b	$8,433 \pm 300$	$9,153 \pm 300$		$10,210 \pm 150$
Hl_2a			$9,150 \pm 220$	
Hl_1b	$8,435 \pm 300$		$9,100 \pm 220$	$11,790 \pm 150$

Using pollen zones as a stratigraphic control in a discussion of lake sediments and radiocarbon dates in South West New Brunswick, Karrow and Anderson (1975) concluded that some of the dates were in error by as much as several thousand years. One example was 4000 years too old. These anomalies were related to the changing chemical environments of small lake basins after deglaciation. Newly deglaciated terrain, it was argued, would be unweathered, and carbonate bedrock, and or carbonate rich glacial deposits would produce carbonate rich sediments. Then later as the vegetational cover developed, and pedogenesis proceeded, more and more organic debris would be contributed to small lakes. This prediction is fulfilled in the lakes under consideration in that the basal sequence consists of marl, or marl-like sediment, overlain by organic mud (*gyttja*) and peat. Pollen analysis shows that the carbonate content of the deposits decrease sharply concomitant with the spread of spruce dominated vegetation. The result of this sequence is that greater amounts of old carbon are likely to be incorporated in the early stages of the lake development, hence radiocarbon dates from such levels are anamalously old when compared with adjacent data.

Anomalous dates of this kind are sometimes disregarded, but sometimes they have been used as a basis for inference about Pleistocene events, as in the case from the Isle of Man, U.K. Radiocarbon dates between $18,900 \pm 330$ and $18,400 \pm 500$ BP were obtained on moss fragments from basal deposits within kettle holes at Glen Ballyre. The dates were used to suggest that deglaciation of the Late Devensian Irish Sea Sheet had occurred by that time (Thomas 1976). At no time was the possibility considered that the dates were apparent ones only, although Shotton (1977b) conceded that such a possibility existed, especially as the moss in question, *Drepanocladus revolvens*, photosynthesized under water.

The foregoing example is commonplace. Ofen inexplicable breaks in sedimentation, or anomalous rates of deposition, appear to be associated with such dates. At Glen Ballyre, for example, dark brown detritus mud, dated at 12,150 BP lies only 0.88 m above the moss layer dated to 18,900 BP, the implication being that only 0.88 m of sediment was deposited in nearly 7000 years, at a time when severe periglacial processes were known to obtain in Britian. Near Corry, in Pennsylvania, a basal marl, dated between 14,000 $\pm$ 350 and 13,000 $\pm$ 300 is overlain immediately by a peat whose base is dated as 9430 $\pm$ 300 BP. The difference in age between the top of the marl and the base of the peat is either attributed to an interval of non-deposition (about 4,000 years), or contamination of the peat from higher levels (Droste 1959). More likely, however, is that the marl dates were two 'old' due to deviations in the carbon reservoir they were associated with in comparison with that of atmospheric carbon.

A final example is that from Jutland, Denmark, where Shotton (1972) was able to demonstrate a stratigraphic succession of dates exhibiting an approximately constant hard water error of 1700 years (Fig. 5-2). He was able to show that, depending on the nature of the sample, the date from a particular horizon varied systematically up-profile. Dates from small twigs and pieces of wood are plotted on the left-hand graph on Figure 5-2. Another set of dates from comparable stratigraphic levels was produced from black mud and moss (lowest sample). The dates not only fail to match the first set, but are inconsistent with the palynological evidence from the site. The age differences are attributed to hard water error. The reliability of the two sets relates to the type of material being assayed; moreover the size of the error may be quantified.

In the cases mentioned above some stratigraphic dating control was available, either from local, or regional data, which suggested that the dates were anomalously old. That is, quite apart from the fact that they should been so regarded in any event due to the nature of the samples. In other cases, however, no such control is available, and it is a triumph for geochemical and hydrological case history investigation that the dates produced appear internally, and externally consistent. An outstanding example of such a procedure is that of the radiocarbon dates produced for high lake levels of pluvial lakes in the western United States.

The stratigraphic and morphologic evidence for high stands of pluvial lake level is considered in Chapter 9 (Morrison 1965). Radiocarbon dating is considered here. The former pluvial lakes occurred in areas of interior drainage which ended in closed basins. Lake levels fluctuated according to climatic change and the sequence of events on stratigraphic grounds appear firmly documented. High lake levels were dated by Broecker and his associates to the period between 22,000 and 8,000 years BP, with the highest occurring at : 17, 15, 12 and 9 ka BP (Broecker and Orr 1958. Broecker and Kaufman 1965). Factors influencing dates on various materials were systematically investigated and much evidence was produced relating to the relative merits of different materials for dating. Broecker and his collaborators used a carbon-14 concentration of 0.94 of that of the atmosphere to calculate ages for Lake Lahontan and Lake Bonneville, which reflected allowances made for the geochemical balances of the lakes and their drainage systems, carbon-14 content of the lakes, and the balance between that brought into the lake, that decaying in the lake, and that which had decayed before being brought into the lake.

In a review of dating problems associated with former lake levels Thurber (1972) concluded that each lake basin presents an unique situation. There are no hard and fast parameters to be applied ubiquitously. But accurate radio-carbon ages may be obtained from such situations only when the environmental background has been exhaustively researched. Otherwise the data are only of limited local value and of apparent age only.

Contamination

Generally speaking organic materials present less of a difficulty than do carbon-ates, for the latter are particularly prone to contamination. In essence contamination of a sample means that it has not remained a closed system since death, due to entry of extraneous carbon. This may occur in three ways : (1) through carbon exchange directly with the atmosphere or with ground water, (2) through the mixing of materials of different ages, (3) by a continuous development of a sample, such as a soil, or by multi-stage development as with some precipitates, or by the recrystallization of the sample - e.g. replacement of original aragonite by calcite.

Contamination by direct exchange with the atmosphere or ground water may take place particularly with the exchange of surface molecules of carbonates. It can also result from careless handling, especially when samples are being washed (Olsson and Eriksson 1965). To minimise this effect, dates on shells are usually made on successive fractions from the outside to the inner part. Anderson (1969) has shown that such contamination is a rapid process and it must be assumed that all carbonates have a contemporary contamination equal to the ratio of surface molecules to the total in the sample. Thus the magnitude of the contamination may be estimated. Studies have demonstrated a generally high percentage of contamination, and it is recommended that dates on carbon-ates in excess of 25,000 years should be regarded as minimum ages only (Thurber 1972). Exchange as a means of bone contamination is also a serious problem (Olsson 1974).

The introduction of humic acid by plant roots or ground water percolation is a particularly serious means of contamination. Pretreatment for the extraction of humic acids is not always entirely successful, and the magnitude of the problem may be appreciated from Figure 5-3, and Table 5-4. Quite small amounts of contamination by modern carbon can produce radiocarbon ages which may be entirely misleading.

After collecting organic and carbonate samples for interglacial deposits in England and Wales Page (1972) obtained a series of radiocarbon dates which compressed that part of the Pleistocene time-scale they represented : for example, the date of the Cromerian Interglacial was 44 to 25.8 ka BP, and the Ipswichian (Last) Interglacial was 19.5 to 18 ka BP. Shotton (1973), however, showed that all these dates were the result of contamination, and, moreover, were obtained from situations where contamination was certain to obtain. Moreover they are contrary to all geological dating evidence.

A large number of samples have provided dates from which a high stand of sea level between 30 and 40,000 BP has been adduced (e.g. Milliman and Emery 1968). It was suggested that this reached at least as high as present-day sea level. This has been a particular source of debate in Australia (Langford-Smith and

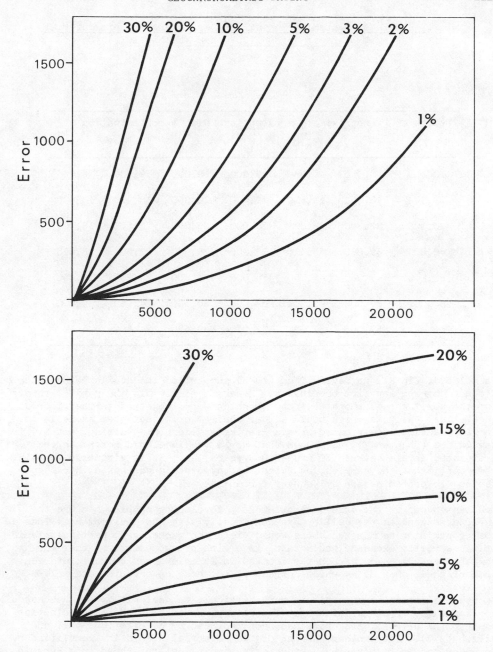

Fig. 5-3.   Effects of % contamination on a sample's apparent carbon-14 age by old carbon (top) and young carbon (below) (after Olsson 1974)

TABLE 5-4   Contamination effects of modern organic material
(after Shotton 1967a and Thurber 1972)

True age	Approximate apparent age with % contamination					
	0.1	0.2	1.0	2.0	5.0	10.0
10,000	10,000	9,960	9,800	9,630	9,100	8,000
20,000	19,900	19,800	19,000	18,300	16,400	13,000
30,000	29,700	29,300	28,000	25,300	21,270	16,000
40,000	39,000	38,110	32,000	29,900	23,800	18,000
50,000	47,000	45,000	34,000	31,650	24,300	18,000
60,000	52,000	49,000	34,000	32,100	24,650	18,000
infinite	57,000	51,000	38,000	32,500	24,800	

Thom 1969). In a long review Thom (1973) summarized the available evidence but
did not come to any firm conclusion. However, nearly all the samples throughout the world, were carbonates, none of which were subjected to the rigorous
geochemical analysis which would either confirm or reject them as reliable.
In the meantime firmly based Uranium-series and carbon-14 dating, notably from
New Guinea (Chapter 7), has shown that world sea level did not reach the
postulated levels between 30 and 40,000 years ago. It is virtually certain,
therefore, that the samples, on which that hypothesis was based, had been
subject to contamination by younger carbon-14.

Problems occur in the case of continuously developing samples because the
carbonates tend to re-equilibriate continually with the environment. Thus a
soil containing carbonate debris would yield ambiguous ages, particularly if
it was wet, for exchange would take place with ground water. Rather more
reliable are dates on carbonates from soils in arid areas because, at least
some of them, show dates which correspond to those on organic material.

Recrystallization of a sample is more difficult to deal with. In the case of
coral samples from New Guinea a laboratory examination was made to detect
whether the original aragonite had been replaced by calcite (Chappell and
Polach 1972). But a clear warning of the potential errors is exemplified by
Broecker and Kaufman's (1965) study of an individual tufa head from Pyramid
Lake, Nevada. The radiocarbon dates speak for themselves :

TABLE 5-5   Radiocarbon dates across a tufa head from
            Pyramid Lake, Nevada, U.S.A. (after Broecker
            and Kaufman 1965)

massive tufa	8,500	OUTER SIDE (youngest)
thin zone of shattered tufa	12,150	
dendritic tufa	12,300 14,500 15,500 15,350	
short thinolite crystals	11,000	
coarse thinolite crystals	11,400	
short thinolite crystals	13,450	INNER SIDE (oldest)

Quaternary deposits in areas of Coal Measures often contain particles of old
carbon or spores. Such inert or 'dead' carbon will result in radiocarbon dates
that are overestimates. This, however, is not a serious source of error as is
shown on Figure 5-3. For example, if 1% of the sample consists of old material,
then the radiocarbon age will be 80 years too great; if 10%, 850 years too
great; and if 50% one half-life too great. Donner and Jungner (1973)
describe an interesting case from south-west Finland, where the incorporation
of redeposited organic material, of Eemian age, gave radiocarbon dates that
are too old for material otherwise known to be of late Weichselian and
Holocene age. In this case pollen and diatom grains, of typical interglacial
aspect, enabled the source of contamination to be identified.

URANIUM SERIES DISEQUILIBRIUM DATING METHODS

There are two long-lived Uranium nuclides, ^{238}U and ^{235}U, which decay though a
series of shortlived products to stable lead. Most of the members of this
decay series cannot be used for dating Pleistocene events because their half-
lives are too short, but some have appropriate half-lives. These are : in the
^{238}U series - ^{234}U (half-life of 250,000), and ^{230}Th (half-life of 75,000);
and in the ^{235}U series - ^{231}Pa (half-life of 32,000). The decay series is
shown in Table 5-6. Uranium decays to a daughter isotope by emitting an alpha
particle from its nucleus. In turn the daughter isotope decays with emission
of an electron, and a series of atoms to which the nucleus decays until stable
lead forms.

TABLE 5-6   Decay series of ^{234}U and ^{235}U

Nuclide	Half-life	Nuclide	Half-life
uranium-238	4.51×10^9 yr	uranium-235	0.713×10^9 yr
	$1\alpha, 2\beta$		$1\alpha, 1\beta$
uranium-234 1α	250×10^3 yr	protactinium-231	32.4×10^3 yr
thorium-230 (*ionium*) 1β	75.2×10^3 yr		$2\alpha, 1\beta$
radium-226 1α	1620 yr	radium-223	11.1 day
radon-222	3.83 day	radon-219 (*actinon*)	3.9 second
$3\alpha, 2\beta$		1α	
		polonium-215	1.8×10^{-3} second
lead-210 2β	22 yr		$2\alpha, 2\beta$
polonium-210 1α	138 day		
lead-206	stable	lead-207	stable

The several methods of application of these isotopes are discussed by Broecker (1965), the essential basis for which can be appreciated by means of an analogy with the dynamic nature of a forest, as presented by Broecker and Bender (1972) :

> In a mature forest trees will be dying and new trees will
> be maturing in their place. The total number of full
> grown trees will be more or less constant. If the forest
> is disturbed, it will eventually return to its mature
> condition. If it is destroyed by fire for example, new
> trees will grow in at a characteristic rate. A botanist
> could estimate the age of the forest by noting the extent
> of its return to the mature state. On the other hand, if
> there are too many trees in the forest, the number will

gradually decrease to the 'mature' amount the forest can support.

In what Broecker and Bender (1972) term a 'mature' rock (>1 Ma) a dynamic condition obtains wherein the dozen or so isotopes formed by radioactive decay are decaying at rates similar to those of their formation from the parent. If this equilibrium is disturbed, through the chemical removal of the products, then re-establishment of the process back to equilibrium is initiated. In the case of ^{230}Th/U and related methods, dating is based on the rate of return of the decay rates of the daughter nuclides to the rates which obtain in a 'mature' or equilibrium system.

Uranium daughter isotopes are virtually absent from water. This means that carbonate organisms and inorganic carbonates will have taken up uranium, but none of its daughter isotopes. Provided that a given sample has remained a closed system, then the extent to which daughter isotopes have reappeared can provide a measure of the age.

The methods are applicable to deep-sea sediments, corals, calcitic molluscs and aragonitic molluscs. All these will have taken up uranium from water, but none of its isotopes. The extent to which ^{230}Th and ^{231}Pa have reappeared in a fossil sample, provides a measure of its age. This, however, is dependent on two major assumptions : (1) that the initial ^{230}Th/^{238}U in the sample is known, and (2) that the system has remained closed. A major source of error with regard to (2) is that diagenetic alteration takes place, particularly among older samples. If undetected this can provide erroneous dates (below). Rigorous selection of samples is, therefore, necessary, both in the field and in the laboratory.

The dating of fossil corals

Highly successful attempts at dating fossil corals have been made. This is of crucial importance in the elucidation of sea level change during the Quaternary (Chapter 7), and the correlation of continental with marine events as revealed by deep-sea core data. Dates may be regarded with a high degree of confidence provided that the following checks have been made (Thurber *et al* (1965) :

(1)   aragonite is used for dating and no crystallization to calcite has occurred

(2)   the uranium content of the coral is the same as its modern counterpart : that is, about 3 parts per million

(3)   the ^{234}U/^{238}U ratio, after decay allowance, was 1.15 : that is, the same as in living corals and in sea water. Such a high value in sea water is due to its influx by rivers (Broecker and Bender 1972)

(4)   the ^{226}Ra/^{230}Th ratio agrees with the ^{230}Th/U age

(5)   little or none of ^{232}Th is present in the sample

(6) ^{230}Th/^{234}U age is the same as the ^{14}C age from the
sample, when it falls into the range of the radio-
carbon dating method (see Chapter 7)

(7)   the ^{230}Th/^{234}U age is consistent with the available
stratigraphic and other assessment as regards its
age.

Changes over time of the various activity ratios are shown in Figure 5-4.
For example, that of ^{234}U/^{238}U starts from 1.15 (its value in sea water) and
decreases to 1.0, which is the equilibrium value found in old rocks. Other
ratios increase from zero to 1.00, which represents the equilibrium value.
Thus, the age of a sample may be read off the graph, which is based on
experimental data : for example, a sample with a ^{230}Th/^{234}U ratio of 0.70 is
125,000 years (Fig. 5-4). Remarkable dates have been obtained from coral
reefs in Barbados (Mesolella *et al* 1969) and the Huon Peninsula of New
Guinea (Chappell 1974. Bloom *et al* 1974) and these are considered in Chapter
7.

Fossil corals may also be dated by the helium-uranium method which is based on
three assumptions : (1) that changes are only due to radioactive decay; (2)
all helium results from the decay of uranium and its daughters; (3) other
than helium lost through a predictable and understood process, known as alpha
stopping (Broecker and Bender 1972), all generated helium is not lost to the
system. Some He/U ages from aragonite molluscs and corals have yielded dates
which are consistent with ^{230}Th dates, and are otherwise in agreement with
stratigraphic assessment.

The dating of fossil molluscs

Unfortunately the dating of fossil molluscs has been unsuccessful despite
several apparently sound applications. When the relevant geochemistry is
considered this lack of success is by no means surprising.

For example, only one fiftieth the amount of uranium found in corals occurs
in molluscs. Furthermore the system does not remain a closed one because
further uptake of uranium occurs during diagenesis : ^{238}U increases by a
factor of 25 during the first 10,000 years after death, and by a further 50%
during the following 100,000 years. It is also known that a process exists
that adds ^{234}U, relative to ^{238}U, for many hundreds of thousands of years.
When checked against independent means of dating, most molluscs do not
exhibit similar ages by ^{230}Th/^{234}U and ^{231}Pa/^{235}U methods, nor are they
consistent with stratigraphic data.

Kaufman, Broecker, Ku and Thurber (1971) wrote a paper with the avowed purpose
of demonstrating that dates obtained on molluscs by this method are unreliable.
They presented data from 400 analyses, from which it was thought no more than
60 samples produced correct ages. All the others displayed variable evidence
of having been open systems. They also criticised an open system model
(Szabo and Rosholt 1961) as being based on unlikely isotopic migration
patterns and on unlikely assumptions. No further progress seems likely until
the open system types of migration in U-series isotopes are understood, for

they are complicated and probably multi-staged (Kaufman *et al* 1971). Some
dates, however, are undeniably reasonable, and tend to confirm predicted ages.
For example, in one of the earliest dating attempts, dates from molluscs and
beach rock from Mediterranean raised beaches yielded ages which were not only
internally consistent, but also confirmed some similarly early dates from
elsewhere, notably on corals. The dates, which occurred in three groups,
were : 75 - 95, 110-140 and >200 ka (Stearns and Thurber 1965). In a later
evaluation Broecker and Bender (1972) suggested that such favourable dates
resulted from the protection of the molluscs within carbonate beach rock. In
other words, the system remained relatively closed.

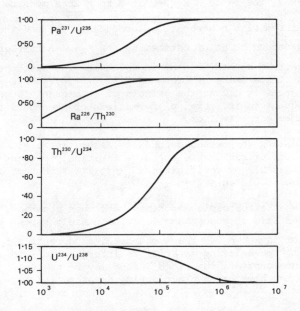

Fig. 5-4.   Changes through time in the ratio of U-series isotopes (from
Broecker and Bender 1972)

POTASSIUM-ARGON AND ARGON-40/ARGON-39 DATING

This method is applicable to volcanic rocks such as lavas and tuffs, thus is restricted to areas of Quaternary vulcanicity. These are usually associated with zones of Cenozoic uplift or lithospheric plate boundaries, such as East Africa, the Rocky Mountains, or regions of localized activity such as Germany.

Potassium occurs in volcanic rocks in amounts up to 10%, consisting of stable ^{39}K, stable ^{41}K, and a small amount (0.012%) of the unstable ^{40}K, which decays to ^{40}Ar and ^{40}Ca. Because ^{40}Ca is already abundant in most rocks, the decay to ^{40}Ar is the one used for dating purposes. Thus with the half-life of ^{40}K known (1.3 x 10^9 years), the ratio ^{40}K ^{40}Ar in a rock is a measure of its age. If the sample is irradiated with neutrons, ^{39}K produces ^{39}Ar, and the ratio $^{40}Ar/^{39}Ar$ provides a second means of estimating the age. A further method, the $^{40}Ar/^{39}Ar$ age spectrum method, is one in which the isotopic ratios are measured at different temperatures (Fitch and Miller 1970). This is useful because in some instances it allows the effects of subsequent 'overprinting' (due to events subsequent to the original formations of the rock) to be evaluated and sometimes corrected for.

As with radiocarbon dating the method depends on assumptions. It is assumed that a negligible amount of radiogenic argon was initially present in the rock, something which is incapable of verification; and, it is further assumed that the sample has remained a closed system.

Uncertainties arise when atmospheric argon may have been absorbed, a difficulty particularly characteristic of young samples. Fortunately by checking the $^{40}Ar/^{36}Ar$ ratio, which is known for the atmosphere, a correction factor may be applied. Occasionally pre-existing material is incorporated into volcanic rocks, thus a certain amount of 'inherited' argon may be present. Further, secondary alteration may have taken place during metasomatism or weathering processes.

Such is the minute quantity of ^{40}Ar in very young rocks that it is extremely difficult to measure. For example, in a rock 10,000 years old containing 2% potassium, argon only occurs to the extent of 10^{-12} (Aitken 1974), hence high instrumental sensitivity is required. The realistic limit of potassium-argon dating is variously estimated : some would restrict it to rocks older than 250,000 years (Shotton 1977a), others to rocks older than 100,000 years (Miller 1972), but in special instances dates of between 6,000 and 10,000 years have been measured (Dalrymple 1968). In general, however, the older the sample, the more reliable the age is likely to be.

As with other forms of radiometric dating it is desirable that the ages produced are capable of independent verification. It is recommended in this instance that reproducibility by the same, and if possible other methods, is attempted from the same geological horizon. A potential pitfall here is that all the samples from a given locality have been similarly 'overprinted' by the effects of subsequent events. Partial overprinting, however, can be investigated by means of $^{40}Ar/^{39}Ar$ spectrum analysis (Fitch 1972).

Characteristic of the method's result has been the production of both
anomalously young and old ages. Those that are too 'young' are the most
common, and usually derive from the loss of radiogenic argon from the system.
This occurs as a result of diffusion, chemical reaction, a change of state,
solution or recrystallization. With care, such uncertainties can be estimated
and allowed for. Anomalously 'old' dates obtain because of the addition of
extraneous radiogenic argon from, for example, the particular mode of formation
of the rock or mineral (Fitch 1972). Excess argon enter minerals by diffusion,
or is retained trapped in occlusions, but fortunately its presence can be
detected by spectrum analysis.

Potassium-argon dating attains maximum usefulness in the Middle and Early
Pleistocene. In areas of Cenozoic activity it has enabled the dating of
glacial advances in the Rocky Mountains (e.g. Richmond 1976) and in South
America extending back into the Pliocene (Mercer 1974). Its value for dating
archaeological fossils is well known, for example the hominid remains, at Lake
Rudolf, which are associated with volcanic rocks, to 2.42 Ma (Fitch *et al*
1976) (Chapter 6). But its role in dating palaeomagnetic reversals in the
earth's magnetic field, as revealed in the stratigraphic record, is of crucial
importance as a means of correlation (below).

FISSION TRACK DATING

Traces of ^{238}U occur in minerals and glasses of volcanic rocks. This decays
by spontaneous fission (explosive division into two fragments), at a rate of
$\sim 10^{-16}$ per year, and causes substantial damage to the lattice by recoiling
fragments. After the damage tracks have been etched, with for example acid,
they are readily visible under an optical microscope. The number of tracks
present is a measure of the age of the sample, and they consist of single
narrow but intense trails of damage between 5 and 20 microns in length. The
absolute rate at which the tracks form is proportional to the ^{238}U content of
the rock (Fleischer and Hart 1972).

The method is particularly useful where volcanic rocks occur and provides
cross-checks on other dating methods : for example, the age of the KBS tuff
from Lake Rudolf is dated as 2.42 Ma (Hurford *et al* 1976), which compares
exactly with the K-Ar age (above). Elsewhere it is invaluable in calibrating
tephrochronological events, as in New Zealand, and the timing of montane
glaciation.

MAGNETOSTRATIGRAPHY

Magnetostratigraphy is a comparatively recent branch of stratigraphy, for
example, The Polarity Time Scale Commission (International Union of Geological
Sciences) was only set up in 1972. This was necessary in a rapidly expanding
field of research not least to define suitable terms in an attempt to avoid
the circular arguments to which the time scale is particularly prone (Watkins
1971. 1972). It is necessary to distinguish magnetostratigraphy clearly from
earlier polarity studies which were compiled from widely scattered data
integrated into a synthetic time-scale. Magnetostratigraphy on the other hand
is based on stratigraphic sequences.

It has long been recognised that the earth's magnetic field is not constant
and has been subject to reversals. For any locality, at any given time, the
magnetic field may be defined by three parameters : (1) declination (D), or

compass variability, which is the angle between magnetic north and geographical
(true) north, (2) dip, or inclination (I), the inclination to the horizontal
of a freely suspended needle, and which varies from 0° at the magnetic equator
to 90° at the magnetic pole, (3) intensity (F), or field strength, which has
a value at the equator one half that at the poles. Magnetic minerals in rocks
show that the orientation and intensity of the earth's magnetic field has
varied. These minerals acquired a remanent magnetism correponding to the
prevailing magnetic field (NRM = natural remanent magnetism). In the case of
volcanic rocks, such as lavas, hard rock solidifies before the temperature
falls below the Currie point of the magnetic minerals. The magnetization
acquired is of thermoremanent origin (TRM). Such volcanic rocks are particularly valuable because reversal boundaries may be dated by the K/Ar method.
This was the earliest basis for erecting a polarity time scale (Cox *et al*
1963). Sedimentary rocks may also preserve a record of palaeomagnetism, and
the alignment of sedimentary grains as they settle through water or water
saturated sediment is known as detrital remanent magnetism (DRM). Magnetostratigraphy may also be successfully applied to deep sea core sediments
(Opdyke 1972).

Uncertainties

Major problems relate to the reliability of the palaeomagnetic signature of
volcanic samples. Both are subject to the effect of post-depositional magnetic
events whereby some of their components may be altered. Laboratory methods
to minimize this effect involve magnetic field demagnetizing treatment which
identifies unstable components and reveals any secondary remanent components.
Reversals in some igneous rocks are mineralogically (chemically) controlled,
and some have been shown to have self reversal mechanisms. It is, therefore,
standard procedure to investigate such a source of high risk in overprinting
on the original polarity. It is also recommended that several K/Ar dates
be obtained from more than one geographically oriented sample to eliminate
dating uncertainties.

Sediments pose greater uncertainties for the original palaeomagnetic signature
may be imperfectly registered due to dynamic distortion by the depositional
process; those of quiet sedimentation seem best suited: e.g. lake clays.
Furthermore post-depositional processes, especially chemical ones, can lead
to the overprinting of the original magnetism. The Polarity Time Scale
Commission recommend that no definitive magnetostratigraphic unit be proposed
unless multiple sampling and laboratory procedures to remove unstable
components and overprinting have taken place. Moreover the units should be
replicated in sequences with different sedimentation rates, lithologies,
depositional and post-depositional environments (Watkins 1976). On ocean
floors difficulties arise through mixing effects which ensure that short time-
scale reversals will not generally be detected. Major events are recorded
although boundaries may be blurred.

Polarity Time-Scale

Table 5-7 outlines the major features of the time scale applicable to the
Quaternary. Bernard Brunhes first recognised that the dipole field of the
earth had been reversed from his studies on French lavas, and later
M. Matuyama, working on lavas in Korea and Japan, concluded that the field was
reversed during the early Quaternary. During the 1950's increasingly
sophisticated instruments for measurement eventually led to the first time-
scale (Cox *et al* 1963) based on 9 scattered data points (For a history of the
developments reference should be made to Watkins (1972) and Dalrymple (1972)).

TABLE 5-7   Polarity Time-Scale for the Quaternary*

Palaeomagnetic Polarity Epoch	Palaeomagnetic Polarity Event	Age	Palaeomagnetic Polarity Excursion
		13.75-12.35 ka	Gothenburg
BRUNHES		20-8.73 ka	Laschamp
NORMAL		30 ka	Lake Mungo
		110 ka	Blake
		——— 700 ka (0.70 Ma) ———	
MATUYAMA	Jaramillo	890-950 ka	
REVERSED		0.89-0.95 Ma	
	Olduvai	1.62-1.83 Ma	
		——— 2.41-2.84 Ma ———	
GAUSS			
NORMAL			

* Ages for epochs and events from LaBrecque' Kent and Canda (1977), and for excursions from appropriate authorities

A *polarity event* (Table 5-7) is a chronologic unit that is characterized by a single geomagnetic polarity that lasted between 10^4 to 10^5 years. These are named after the localities where they were first discovered.

A *polarity epoch* is a chronologic unit that lasted much longer than an event. They may have lasted 10^5 or 10^6 years and are named after pioneers in the field.

Shorter polarity events, lasting perhaps 10^2 to 10^3 years have been termed: 'short event', 'excursion', 'departure' or 'flip'. It is now recommended that these be called *polarity excursions*, named after localities, and defined as:

> a sequence of virtual geomagnetic poles which may reach
> intermediate latitudes and which may extend beyond 135°
> of latitude from the pole, for a short interval of time,
> before returning to the original polarity (Watkins 1976).

The equivocation in the definition arises because it is by no means clear whether these represent global events due to reversal of the dipole field.

Table 5-7 shows that the Quaternary includes the Brunhes Polarity Epoch, from ca 700 ka to the present, most of the Matuyama Reversed Polarity Epoch, and depending on definition of the base of the Pleistocene (Chapter 10) possibly part of the Gauss Normal Polarity Epoch. The major Polarity Events are the Jaramillo Normal Event, named from the Valles Caldera in New Mexico, and the Olduvai Normal Event, named from Olduvai Gorge, East Africa. Formerly a third event was recognised during the Matuyama Epoch, the Gilsa Event, but subsequently it was shown that the Olduvai Event at its type locality included in its range that of the Gilsa, hence they are the same (Gromme and Hay 1971). Also shown on Table 5-7 are some of the polarity excursions, although they have yet to be demonstrated on a global basis.

Several problems arise in dating and correlating magnetic polarity periods. Olausson and Svenonius (1975) have argued that the existing time-scale can only be provisional because possibilities exist for discovering more events. For example, information is sparse between 300 and 400 ka, and before 1.5 Ma data is far from unequivocal. On theoretical grounds they showed that palaeomagnetism could have changed during every stadial.

Coupled with the above is that magnetostratigraphy relies heavily on 'count from the top' procedures (Kukla and Nakagawa 1977), thus substantial errors in dating can arise. This has been the case in the past, for example, when the Jaramillo Event was recognised in 1966 it meant that the first normal event below the Brunhes/Matuyama boundary could no longer be termed Gilsa or Olduvai. The fact is that in deep sea cores such boundaries are not dated, including the Brunhess-Matuyama boundary. At Santerno in Italy the top of Early Pleistocene deposits has been correlated with the Olduvai Event by one authority, but another places them in the Brunhes Epoch (Kukla and Nakagawa 1977). It is desirable to utilised every available item of evidence associated with such magnetic boundaries. Increasingly, therefore, other stratigraphic control is being brought to bear on magnetostratigraphy. The Brunhes-Matuyama boundary is related to a fission-track date on pumice at Wanganui, New Zealand (Seward 1974), while several similar dates between 3.10 and 0.38 Ma enable close control on the magnetic stratigraphy of Early Pleistocene marine deposits in that area. A particular good example of multiple means of supporting data is that from the Kinki district of Japan, where a sequence of marine and freshwater deposits are interbedded with tephra. The sequence covers the period from mid-Brunhes to Gilbert Polarity Epochs, and contains five fission track dates, and sequences of sea level, climatic and vegetational change (Maenaka, Yokoyama and Ishida 1977).

It is doubtful if short term variations revealed by the study of late
Pleistocene and Holocene sediments will ever be capable of global applicability. But for correlation on local or provincial scales they may have
value. Such studies include those on glacial clays in Canada and Sweden
(Morner and Lanser 1974); lake deposits, largely of Holocene age in Lakes
Windermere and Michigan (Creer, Gross and Lineback 1976); and glacial
deposits in Washington (Easterbrook and Othberg 1976).

AMINO ACID DIAGENESIS

Amino acids, the basis of protein, exist in organisms. During life they are
bonded in the protein, but after death such bonds break down, thus releasing
the amino acids. As well as releasing free amino acids, those with 'D'
isonomers will convert from 'L' to 'D', an interconversion called racemization.
The rate of breakdown as well as racemization is temperature dependent, hence
for chronological purposes a uniform temperature over time is necessary.
Given uniform temperature, however, the variation in the ratio of certain
amino acids (Hare and Taylor 1970) has been used as a measure of relative
age. Amino acid ratios if calibrated by radiocarbon or other means of dating
can in some circumstances provide a tool of some importance (Schroeder and
Bada 1976). Note, however, that a temperature uncertainty of only $\pm$ 1% can
lead to errors of about 20%.

Deep cave environments would seem to provide suitable instances where temperature variation through time has been more or less constant. Muleta Cave,
Majorca, was investigated, and amino acid ratios produced for bone samples.
These data compared favourably with dates obtained by radiocarbon and
$U/^{230}Th$ (Turekian and Bada 1972). They suggested that dating back to
200,000 years was feasible. Using fossil bivalves from Pleistocene deposits
along the Clyde Cliffs, Baffin Island, Andrews and Miller (1976) were able to
use amino acid ratios for stratigraphic correlation.

The fundamental constraint of the method, however, is the requirement for
constant temperature conditions, but future work may allow correction factors
for this to be used, particularly if good climatic data is available for the
site. Otherwise it may be confined to localised correlation, sometimes
tagged with radiometric ages, although Andrews and Miller (1976) believe that
it is potentially applicable along most of the 1200 km of the east Baffin
Island coastline.

GLACIAL VARVES

Baron de Geer commenced his monumental work on varves in Sweden in 1878.
Varves are formed when glacial meltwater is discharged into still water,
usually a proglacial lake. During the summer, meltwater carries a suspended
load of sand and clay, and on sedimentation, the coarse grains settle first
and the fine ones later. The result is a regular sequence of annual deposits,
clearly visible, for the change from coarse to finer material is indicated by
a change in colour, from light to dark. Thicknesses vary from a few mm to
several cms, and are determined largely by climate: e.g. a hot summer will
produce enhanced melting therefore thicker varves. By cross-correlating the
upper layers at one site with the lower ones at another, a chronology can be
built up.

de Geer built up a chronology which related to the retreat of the last ice
sheet in Sweden from the Scanian moraines in the south to Lake Ragunda in the

north. He took his zero point as the bipartition of the ice sheet, which occurred close to Lake Ragunda, estimated at 6,923 B.C. Nowadays, however, the zero date is linked with the drainage of the Baltic Ice Lake for with the opening of the dam, lake level fell and subsequent varves were both thicker and darker in colour. This event, variously estimated, was dated by Morner (1969) to 10,163 BP from varves, and 10,000 to 9,950 BP by radiocarbon. In de Geer's terminology it marks the final retreat from the Younger Dryas moraines (Chapter 2) - the Finniglacial retreat, as opposed to the earlier Gotiglacial retreat from Scania, and the yet earlier Daniglacial retreat from Jutland. Retreat took 3600 years (de Geer 1940) was was related to a scale of calendar years through linking the glacial with estuarine varves; the upper most of the latter was dated to 900 A.D. (1,050 BP), which in turn dates the commencement of retreat at 10,500 B.C. (11, 450 BP), and its end at 6,900 B.C. (8,850 BP). More recently the date of the uppermost river varve has been assessed at 700 A.D. (1,250 BP).

The chronology of de Geer has been linked to a calendar scale by several workers (e.g. Jarnefors 1963), but in a review of all the possible errors Fromm (1970) showed that it was accurate only to within $+ 200 ^{+ 475}_{- 300}$ years at about 12,000 BP, and less accurate at younger dates. The chronology has also been linked with Pollen Zones, and with radiocarbon dating as calibrated by bristlecone pine (Tauber 1970). But Lundquist (1975) shows that it is premature to make such comparisons in central Sweden, and concludes that the de Geer time-scale is even less reliable than has become accepted in recent years. More field date is required in critical areas, especially between Gavle and Sundsvall on the Swedish Baltic coast. South of Gavle the area has not yet been connected with the chronology so that the maximum error could be in excess of 675 years.

Less success has obtained in varve chronology in North America, which is complicated by gaps in the record, and differences in rates of advance and retreat between different ice lobes (Anteves 1955). Moreover there is a discrepancy between varve chronology and radiocarbon dates (Flint 1971). In an interesting study on varves from the Lake of the Clouds, Minnesota, Stuiver (1970) made radiocarbon measurements on a core with 9500 laminations. The results, however, indicated that there was a carbon-14 excess back to 9950 BP amounting to 10%, and they do not agree with those from Scandinavia (Tauber 1970).

NON-GLACIAL ANNUAL DEPOSITS

Non-glacial annual deposits have long been known from pre-Pleistocene periods (Zeuner 1946), but they are also known from interglacial and Holocene lakes in Europe. They are of some importance because, in the absence of any radiometric dating, they allow an estimate of the duration of interglacials. Turner (1975) has reviewed the data from Europe and concludes that five conditions in particular are conducive to their formation: (1) small, steep-sided, deep lakes; (2) lakes well sheltered from wind, which would cause waves and hence disturbance of the sediment; (3) lakes with small tributary streams; (4) a sharp seasonal contrast between winter and summer, as would, for example, occur in more continental environments; (5) a high organic productivity in surface waters.

In detail each pair of laminae consists of a darker layer richer in organic material, and a paler one sometimes consisting of calcium carbonate. In

thickness they range from 1 to 2 mm, or may be thinner. Sedimentological
investigations have confirmed that they are seasonal deposits, and in one
case, from the Schleinsee (Müller 1971), which is of Holocene age, the darker
layers contained pollen grains of plants which flower during the autumn and
early spring, whereas the lighter coloured layer contained the pollen of summer
flowering plants. Thus it can be seen that the darker layers represent the
winter half of the year, and the lighter, carbonate rich layers, the summer
half.

These annual deposits are known from several interglacial deposits in Europe.
At Bilshausen Müller was able to relate the number of laminations to pollen
zones of the Ruhme Interglacial (Table 5-8). By adding to these estimates
for laminations missing because of incomplete cores, or because they were too
fine for counting, he suggested that the interglacial lasted between 28,000
and 36,000 years. For the Holstein interglacial laminated deposits have been
investigated at two sites in Germany by Müller, at Hetendorf and Munster-
Breloh, and in England at Marks Tey (Turner 1970. Shackleton and Turner 1967).
At Marks Tey it was not possible to make accurate counts of the laminations
throughout the deposits, and some estimated figures were necessary. However,
it was suggested that the Hoxnian Interglacial represented there lasted for
30,000 to 50,000 years, an estimate subsequently revised to 20,000, or at the
most 25,000 years (Turner 1975).. Finally, the Eemian interglacial diatomite
deposits at Bispingen was investigated by Müller who, by using comparative
data, was able to estimate its duration as about 10,000 $\pm$ 1,000 years.

Table 5-8   Pollen zones and laminations of the Ruhme
             Interglacial at Bilshausen (after Müller
             1965)

Picea-Betula-Pinus zone	ca. 2,200 years
Picea zone	ca. 1,000 years
Quercus-Abies-Carpinus zone	ca. 9,300 years
	(but because of disturbances might be 6,000-14,000)
Abies-Carpinus-Quercetum mixtum zone	ca. 1,200 years
Quercetum mixtum zone	ca. 1,800 years
Betula-Pinus expansion	ca. 400 years
Picea-Abies-Quercetum mixtum zone	ca. 4,000 years
Late *Ulmus-Picea-Pinus* zone	ca. 6,000 years
Early *Ulmus-Picea-Pinus* zone	ca. 4,500 years

DENDROCHRONOLOGY

The principles of tree-ring chronology are well understood (Fritts 1965).
Annual growth rings are composed of wood cells with large lumens which grow
in the spring, and smaller cells in summer and autumn. Wet summers produce
thicker rings, but conditions of stress, such as droughts, may eliminate a
ring altogether. Hence comparison of present day stress conditions and
ensuing tree ring characteristics, with the record of former seasons is
necessary. Environmental conditions clearly condition tree ring characteristics so that correlation is only really feasible at a provincial scale. By

overlapping tree rings from older to younger trees a chronology which extends
for almost all of the Holocene has been erected in western America. Calibration by carbon-14, however, has demonstrated serious divergences (above), but
within the historic period it continues to be a useful tool.

Recent work has enabled detailed climatic parametrs to be established
from tree ring characteristics, such as precipitation values, temperature
variations, and data relating to air mass climatology, including periods of
storminess (Blasing and Fritts 1976).

Chapter 6
THE FOSSIL RECORD

Some of the constraints in utilising the fossil record for subdivision of the Quaternary have already been hinted at, or explicitly made (Chapters 1, 2 and 4). Principally they derive from the relative brevity of the Period which did not result in appreciable evolution in most fossil groups, although a high vetebrate evolution, notably in man, occurred. And also because the repeated alternations of cooling and warming tended to reproduce essentially similar constellations of species, particularly during interglacials. Such facies, faunas and floras, occasionally characterized by extinction or first appearances, were subject to migration at finite rates dependent on the constraints of environment.

With the new facts of climatic change available from the deep-sea record (Chapter 3) it is necessary to re-evaluate much of the traditional biogeographical dogma on the Quaternary. For example, biogeographers have traditionally tended to regard glacials and interglacials as being of sub-equal length. Now that it is known that interglacials, at least in mid-latitudes, only constitute 10% of the time, it follows that the glacial vegetational state, with of course variations within it, should be regarded as the norm. The traditional concept of stable vegetational belts moving alternately southwards and northwards in response to climatic change also requires modification. According to this model, for example, the Mediterranean lands consisted of deciduous woodland during the glacials. Yet pollen analytical evidence shows quite clearly that this was not the case, and instead, widespread aridity obtained. A further difficulty arises because the uniformitarian method is often limited in value for fossil assemblages demonstrate the former existence of environments without modern analogues. Major biogeographical patterns are by no means sacrosanct.

The difficulties increase the farther back investigations go. But whereas the use of biological evidence for purposes of correlation and dating is not as powerful as was once thought, a veritable tidal wave of investigations has provided powerful tools for palaeoecological reconstruction.

FLORA

Inference from fossil flora provides data on environment (Table 6-1), and by secondary inference, climate also. Much depends on knowledge of modern ecosystems, taphonomy of the fossil assemblages and taxonomy for successful inference and reconstruction. Plants macrofossils consist if wood fragments, leaves, seeds and fruits, and usualy reflect local environments of derived from autochthonous deposits. But mixed assemblages occur in allochthonous deposits and interpretation will vary according to the transporting process. Microfossils on the other hand indicate both local and regional vegetation in that transported by air they may be deposited widely.

TABLE 6-1   Specific climatic indicators : Holstein inter-
             glacial and present day compared (from Frenzel
             1973)

| | Mean temperature | | | Annual Precipi- |
	January °C	July °C	Annual °C	tation (mm)
British Isles				
Vitis sylvestris	=	+2	+1	=
Azolla filiculoides	=	+2	=	?
Denmark				
Vitis sylvestris	+1 to 2	+3	+2 to 3	=
Buxus sempervirens	+1 to 2	+1	+1 to 2	=
Trapa natans	?	+2	?	?
Northwestern Germany and the Netherlands				
Vitis sylvestris	=	+2	+1	=
Staphylea pinnata	=	+2	=	=
Poland				
Vitis sylvestris	+2	=	+1 to 2	=
Azolla filiculoides	+3	=	+1	?
Fagus silvatica	+3	=	=	=
Tilia platyphyllos	+1 to 2	=	+1	+50
Ilex aquifolium	+3 to 4	=	=	=
Southern Lithuania				
Tilia platphyllos	+5 to 6	+3	+4	=
T. tomentosa	+6	+5	+7	=
Carya sp.	=	+3	+1	+150
Central Russia				
Azolla filiculoides	+8	=	+5	?
Ilex aquifolium	+9 to 10	=	+3	=
Tilia platyphyllos	+6 to 9	+2	+4 to 6	+100
Fagus silvatica	+10	=	+5 to 6	=
Trapa natans	?	=	?	?
Taxus baccata	+3 to 4	=	+3	=
Middle reaches of the Volga				
Azolla filiculoides	+11	=	+3 to 4	=
Ilex aquifolium	+14	=	+6?	+300
Lower reaches of the Irtys				
Azolla filiculoides	+24	+1	+12	?
Tomsk Romosk				
Azolla filiculoides	+20 to 22	+1	+11	?

POLLEN ANALYSIS

Method and presentation

Identification based on pollen morphology, isolation of pollen grains from
various sediments, preservation of fossils etc are considered by Faegri and
Iversen (1964) and West (1971). The principal environments where pollen
grains accumulate today are those of wet peat surfaces, and in lakes. But
pollen has been recovered from a variety of sediments, including cave earths,
soils and even from deep-sea cores (e.g. Heusser and Balsam 1977).

Data from pollen analytical investigation is presented in tabular form, or
more usually as a pollen diagram (Fig. 6-1). Three kinds of such diagram
occur. Firstly, the percentage of each taxon in a sample, of all the
communities, or of the tree community only, is shown at given levels in a
deposit : that is, as a plot through time (Fig. 6-1). Secondly, generalized
diagrams may be constructed to summarize the history, or to present an
opinion as to the nature of vegetational change for a given period (Fig. 6-3).
Thirdly, an absolute pollen diagram may be constructed provided that adequate
dating calibration exists. This kind of diagram shows the number of pollen
grains of a given taxon that are deposited on a given surface area per year
(Davis and Deevey 1964. Davis 1969. Pennington and Bonny 1971). It has
certain advantages over the precentage frequency diagram in that the latter is
clearly capable of masking important changes in the vegetation over time. It
reflects changes in input and hence is unaffected by changes in the relative
frequency of other taxons. Figure 6-3 readily illustrates the differences
between percentage and absolute pollen diagrams; both are based on the same
data from Rogers Lake, Connecticut, New England (Davis 1969). The record
spans some 14,000 years, from deglaciation of the last ice sheet through to
the present day. The difference between the two diagrams is greatest during
the first 6,000 years when rapid changes in productivity obtained. The
absolute diagram, therefore, is not only more informative, but allows some
safeguard against erroneous inference.

Sources of uncertainty

These largely concern knowledge of present environments and modes of pollen
dispersal. The relationship between present day vegetation and pollen
sedimentation has been investigated exhaustively in eastern North America,
where the vegetation is less disturbed than it is in Europe (Davis and Webb
1975). Some anomalies occur : there is, for example, no modern representative
for the late-glacial spruce zone of the Middle West, and many other fossil
assemblages lack present day counterparts.

Differential pollen productivity is a variable for which some data is
available. Iversen indicated three main groups : High Producers such as
Pinus, Betula, Alnus and *Corylus*, Moderate Producers, such as *Picea, Quercus,
Fagus* and *Tilia*, and Low Producers such as *Ilex, Viscum* and *Lonicera*. Total
production also depends on the number of individuals, frequency of flowering
and also pollen grain characteristics. Each pollen containing sediment has
usually accumulated under unique environmental conditions which must be
evaluated. Water dispersal depends on local hydrologic conditions. Marine
dispersal encourages over-representation of pine, for its winged grains float
radily. Air dispersal involves local (<50 km), regional (50-100 km) and long
distance transport components (>100 km) (West 1971). Long distance transportation predominates in the Arctic, local and regional components in the Tundra
though the long distance component is high, but in forested lands regional

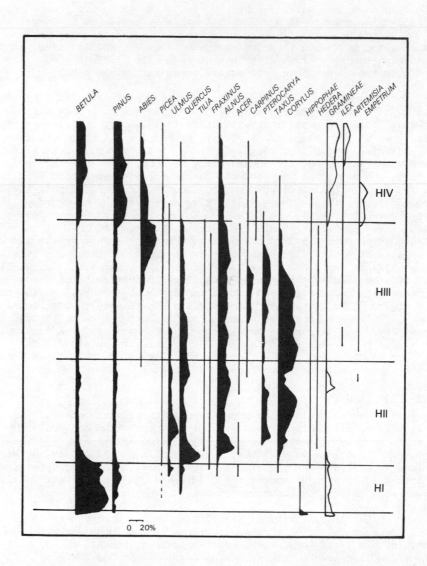

Fig. 6-1.   Pollen diagram for the Hoxnian interglacial from
Marks Tey, Essex, England (after Turner 1970)

production is high. Differential deposition also occurs (Davis *et al* 1971),
due to grain characteristics, recirculation, and the effects of animal mixing.

Interglacials

The definition on an interglacial on palaeobotanical grounds has already been
discussed (Chapter 4). In North America interglacial floras are comparatively
poorly known, whereas their study has been central to Quaternary investigations
in northern Europe. A strong parallelism in vegetational development, with,
of course, local differences, led Iversen (1958) to postulate that this
reflected an equally strong parallelism in climatic development throughout
most of northern Europe. This parallelism was also considered by Andersen
(1966) and their combined views are illustrated in Figure 6-4. It shows a
procession of vegetational events controlled principally by climate, although
in the later stages progressive leaching of soils, leading to podsolization,
has the effects of increasing coniferous tree cover at the expense of the
deciduous one.

Their scheme reflected the established vegetational development of interglacials in Europe (e.g. Table 2-4). Each interglacial, however, was zoned
separately by individual investigators, thus making comparison difficult. In
an attempt to standardize procedure and enhance comparison Turner and West
(1968) proposed that a standard zonation scheme, based on the views of Iversen
and Andersen (Fig. 6-4), be applied. They recognised four zones :

> Zone VI Post temperate zone : return to boreal trees, *Betula*
> and *Pinus*, forest thinning and development of open communities.
> Ericaceous heaths often continue into the following early
> glacials.

> Zone III Late temperature zone (oligocratic): expansion of
> late immigrants such as *Carpinus* and *Abies* and *Picea* with
> accompanying decline of mixed oak forest partly due to
> deteriorating soil conditions.

> Zone II Early temperate zone (mesocratic) : development of
> mixed oak forest with *Quercus, Ulmus, Fraxinus* and *Corylus*
> on rich soils.

> Zone I Pre-temperate zone : development of forest vegetation
> following a late-glacial phase. *Betula* and *Pinus* are
> characteristic trees, with light demanding herbs and shrubs.

This scheme enables ready subdivision of interglacial pollen diagrams (Fig.
6-1), and also allows convenient study of the behaviour of certain taxons in
relation to zonal boundaries for different interglacials (Fig. 6-5), as well
as offering a means of correlation. There are, however, certain disadvantages.
In the first case it could be argued that the subdivisions should not be
given climatic names, for in reality they are no more than pollen assemblage
biozones. A further difficulty arises because they are based on facies
floras, hence are inherently diachronous, and unsuitable for long distance
correlation. They also carry with them the danger that, just as with classical
models of the Pleistocene, they may be taken, and accepted as absolute standards. For example, they are based on a concept of unidirectional climatic
change, whereas data from the north Atlantic would suggest that climatic

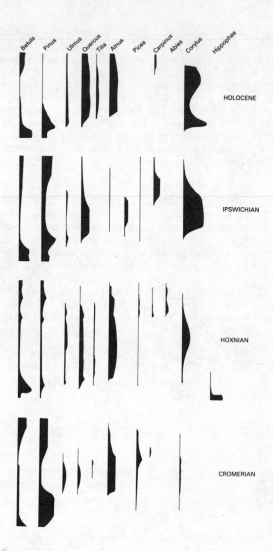

Fig. 6-2.   Summary pollen diagrams for four British interglacials. The relative proportions of different taxons in
each interglacial is used as a means of differentiation
(based on Godwin 1956 and West 1963)

fluctuations occurred in, at least, the last interglacial (below). It is also
likely that fragmentary data will be fitted into the apparent security of
their pigeon-holes and the 'reinforcement syndrome' (Watkins 1971) will start
to operate. For example, the Ipswichian interglacial of England (Fig. 6-6)
is not known to be fully represented by all four zones in continuous sequence
at any locality. Localised fragmentary data may well be fitted into the
scheme whereas in reality they belong to different interglacials. Such has
been suggested to be the case in south-east England, and it may be that three
separate interglacials are involved in what is conventionally called the
Ipswichian (below).

Other than within the range of radiometric dating it is not possible to
establish realistic chronozones based directly on assemblage biozones for a
given interglacial. But a number of chronozones, based on carbon-14 dating,
have been proposed for the Holocene of Northern Europe (Mangerud *et al* 1974).
These are founded on the classical Blytt-Sernander periods which, unfortunately,
have been used with varying meanings. This new proposal, however, allows
comparison to be made with a standard (Table 6-2). For example, Pennington

TABLE 6-2   Subdivision of Flandrian chronozones (after
            Mangerud *et al* 1974)

Chronozone	Subdivision	Radiocarbon years B.P.
Subatlantic	Late	
	Middle	1000
	Early	2000
Subboreal	Late	2500
	Middle	3000
	Early	4000
Atlantic	Late	5000
	Middle	6000
	Early	7000
Boreal	Late	8000
	Early	8500
Preboreal	Late	9000
	Early	9500
		10000

(1975) compared two percentage and absolute pollen diagrams from Blelham Bog,
north-west England, and Cam Loch, north-west Scotland, with the scheme of
Mangerud *et al* (1974). She showed that the chronozone boundaries coincided
with pollen assemblage zone boundaries, but that vegetational differences
occurred between western Britain and Scandinavia. Thus correlation is not
possible by direct comparison of vegetational histories, and only feasible
by chronozone correlation.

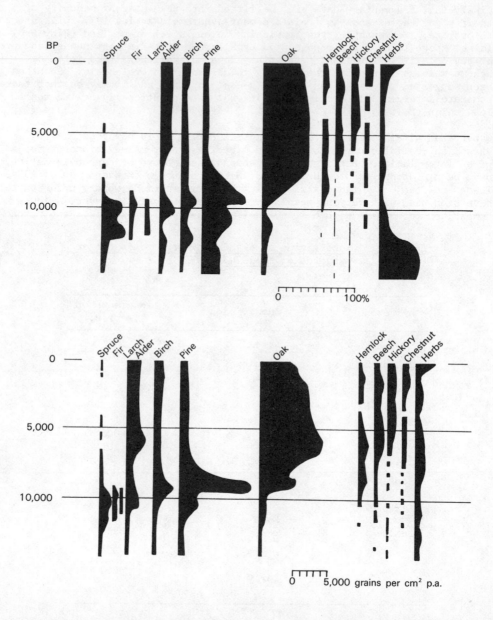

Fig. 6-3.   Percentage (top) and Absolute Pollen (bottom) diagrams from Rogers Lake, Connecticut, U.S.A. (after M.B. Davis 1969)

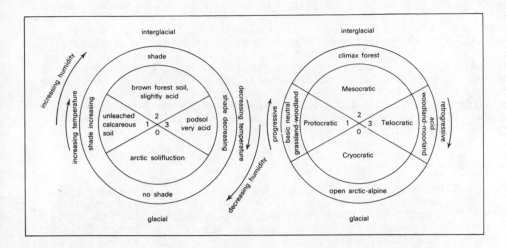

Fig. 6-4.   The Interglacial Cycle (after Iverson 1958 and Andersen 1966)

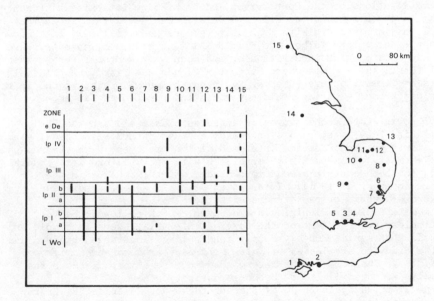

Fig. 6-5.   Biozones of the Ipswichian Interglacial. 1 : Stone; 2 : Selsey; 3 : Ilford; 4 : Aveley; 5 : Trafalgar Square; 6 : Bobbitshole, Ipswich (STRATOTYPE); 7 : Stutton; 8 : Wortwell; 9 : Cambridge; 10 : Wretton; 11 : Beetley; 12 : Swanton Morley; 13 : Mundesley; 14 : Austerfield; 15 : Hutton Henry (after Phillips 1974)

Reviewing the evidence for interglacials based on pollen evidence in Europe, Turner (1975) recognised six since the Menapian cold stage of the Netherlands. Three were pre-Elsterian (Chapter 2), the so-called 'Cromerian Complex', but correlation of these, for example as between Britain and the continent, is not possible. Between the Holstein and Eemian interglacials, but prior to the main Saale glacial advance, occurs the Dömnitz interglacial of northern Germany. A feature of some significance is the recognition of the Brunhes-Matuyama boundary between Interglacial I and part of Interglacial II of the 'Cromerian Complex' (Zagwijn, van Montfrans and Zandstra 1971). Kukla (1975), however, recognised eight interglacials of Brunhes age : three 'Cromerian', Holstein, Dömnitzian, and three that have been incorrectly labelled 'Eemian' in the past (see also Chapter 9).

Interstadials

Interstadials are episodes of climatic amelioration during glacial or cold stages. In some instances they are subject to precise palaeobotanical definition (Chapter 4). Those preceding the Last Interglacial are invariably subject to some uncertainty as they cannot be dated. But for interstadials of the Last Glaciation, a combination of stratigraphic and radiocarbon data allows reasonable dating (Fig. 2-10), although with dates greater than 50 ka a measure of doubt obtains and they should be regarded as minimum ages only.

From investigations of the Port Talbot and Plum Point interstadial (Chapter 2) based on lake deposits in the Erie and Ontario Basins, it was shown that Port Talbot I was warm and dry with mean July temperatures of 15-21° with *Pinus* and *Quercus* present. Port Talbot and Plum Point interstadials, however, were cooler and moister with forest tundra, and mean July temperatures between 10 and 15° (Berti 1975). Elsewhere Watts (1973) has described interstadial deposits of Farmdalian age (Chapter 2) in north west Georgia, which, in the absence of carbon-14 dating could have been interpreted as Sangamon.

Those of north west Europe can be regarded as classical. In the Early Glacial (Fig. 2-10) they record increases in temperature with boreal elements in the Amersfoort and Odderade interstadials but more so in the Brørup which was the warmest. This possibly correlates with the Chelford interstadial (below), for both contain *Picea omorikoides*, which is absent in the Eemian (van der Hammen *et al* 1971). Intervening stadials were characterised by appearance of tundra or park tundra with the herb group widespread. Inter-stadials of the Middle Pleniglacial, however, which followed the very cold vegetationless polar desert of the Lower Pleniglacial, were less cold and more humid than earlier ones. Mostly tundra-like landscapes obtained and among the evidence are leaves of *Salix polaris, herbacea* and *Saxifraga oppostifolia*. The most recent dates, however, are either close to the ordinarily accepted limit of carbon-14 dating, or beyond it (all ka BP): Denekamp ca. 30, Hengelo 36-40.5, Moershoofd 43-50, Odderade 58, Brørup 58.75 - 63 and Amersfoort 63-68. The last three can only be regarded as minima, as can dates on Canadian interstadials in excess of 50,000 years.

MAMMALIAN FAUNAS

The fossil record shows that Early Pleistocene climates had comparatively little effect on the fauna of low latitudes; in Africa great stability obtained (Cooke 1972). But in higher latitudes a gradual disappearance of archaic forest species occurred and ones better adapted to cold appeared (Table 6-3). The Plio-Pleistocene transitional fauns of North America (Blancan mammal age), like those of Europe (Villafranchian) are usually

TABLE 6-3   Faunal changes in Central Europe (from Frenzel 1973)

	1	2	3	4	5	6	7	8	9	10
Equus stenois	+	-	-	-	-	-	-	-	-	-
Archidiskodon planifrons	+	-	-	-	-	-	-	-	-	-
Cervus falconeri	+	-	-	-	-	-	-	-	-	-
Anacus arvenesis	+	+	-	-	+?	-	-	-	-	-
Dicerorhinus etruscus	+?	+	+?	-	+	-	-	-	-	-
Equus robustus	+	+	-	-	-	-	-	-	-	-
Leptobos elatus	+?	+	-	-	-	-	-	-	-	-
Ursus etruscus	+?	+	-	-	-	-	-	-	-	-
Crocuta perrieri	+?	+	-	-	-	-	-	-	-	-
Sus strocci	-	+	-	-	-	-	-	-	-	-
Cervus dicranius	-	+	-	-	-	-	-	-	-	-
C. rhenanus	-	+	-	-	-	-	-	-	-	-
Eucladoceros tegulensis	-	+	-	-	-	-	-	-	-	-
Pannonictis pliocaenica	-	+	-	-	-	-	-	-	-	-
Dicerorhinus kirchbergensis	-	+	+?	-	+	-	+	-	+	-
Trogontherium biosvilletti	-	+	+?	-	+	-	-	-	-	-
Macaca sp.	-	+	+?	-	+	-	+	-	-	-
Hippopotamus cf. *amphibius*	-	+	-	-	+	-	+	-	+	-
Archidiskodon meridionali	-	+	+?	-	+	-	-	-	-	-
Equus Caballus	-	-	+?	-	-	+	+	-	-	+
E. süssenbornesis	-	-	-	-	+	+	-	-	-	-
Cervus elaphus	-	-	-	-	+	+?	+	+	+	+
Palaeoloxodon antiquus	-	-	-	-	+	-	+	-	+	-
Mammontheus trogontherii	-	-	-	-	+?	+	+	+	+?	+?
Sus scorfa	-	-	-	-	+?	-	+	-	+	+
Bos primigenius	-	-	-	-	-	-	+	+	-	+
Ursus spelaeus	-	-	-	-	-	-	-	+	+	+
Alces ales	-	-	-	-	-	-	-	-	+	+
Cervus (Megaceros) giganteus	-	-	-	-	-	-	+[1]	+?	+	+
Ranifer tarandus	-	-	-	-	-	+	-	+	-	+
Lemmus sp.	-	-	-	+	-	+?	-	+	-	+
Coelodonta antiqitatis	-	-	-	-	-	+	-	+	-	+
Mammontheus primigenius	-	-	-	-	-	-	-	+	(+)	+
Bisonpriscus	-	-	-	-	-	-	-	+	+	+
Saiga tatarica	-	-	-	-	-	-	-	+	-	+
Dicrostonyx sp.	-	-	-	+	-	+?	-	+	-	+
Alopex lagopus	-	-	-	-	-	-	-	+?	-	+
Lepus timidus	-	-	-	-	-	-	-	+?	-	+
Microtus gregalis	-	-	-	-	-	-	-	-	-	+
M. nivalis	-	-	-	-	-	-	-	-	-	+
Equus hydruntinus	-	-	-	-	-	-	+	-	+

1 Pretiglian (cold); 2 Tiglian (warm); 3 Eburonian (cold) and Waalian (warm); 4 Menapian (cold); 5 Cromerian (warm); 6 Elster cold; 7 Holsteinian (warm); 8 Saalian (cold); 9 Eemian (warm); 10 Weichselian (cold).

heralded by cold forms. In Europe, for example, by true cattle (*Leptobos*)
horse (*Equus*) and elephant (Kurten 1968), but with persisting archaic forms
such as tapir and *Archidiskodon*.

A strong movement to modern forms occurred during the Middle and Late
Pleistocene in higher latitudes, as well as by adaptation to rapidly changing
climates. But in low latitudes no more than a minor and gradual shift to
modern forms took place. In Europe 20 to 30% of Middle Pleistocene mammals may
be assigned to living species (Kurten 1968). In North American the best
sequences come from the central Great Plains (Hibbard *et al* 1965). There the
Blancan-Irvington mammal age boundary coincides with the extinction of
several Blancan forms, such as the three toed horse (*Nannipus*) and hyena-like
dog (*Borophagus*), together with the appearance of new forms such as Mammouth
and hare (*Lepus*). During the glacials northern forms migrated as far south
as Texas, but assemblages with salamanders and tortoise indicate interglacial
conditions. Immigrants from Asia generally arrived during glacials when sea
level would have uncovered the Bering Land Bridge. The arrival of Bison
(Illinoian) for example, marks the start of the Rancholabrean mammal age.

Problems in palaeoecological interpretation of fossil assemblages are
discussed by Hibbard *et al* (1965) and reviewed by Lundelius (1976) and Kurten
(1968). There is no doubt that considerable changes in tolerance and
functional adaptation occurred. Whereas Zeuner (1959) outlined typical
biotypes for the European Pleistocene, mixed assemblages occur time and again.
Most continents have species that are now allopatric. Such assemblages have
no modern analogues so that uniformitarian principles cannot be applied
easily.

Witthin broad limits some scope for dating and correlation exists.
Unfortunately the evolutionary lineages of only a few groups are known.
Notable among these are European bears and mammoths (Kurten 1968), while
rodent faunas also exhibit marked evolutionary trends (Sutcliffe and Kowlaski
1976). Attempts at dating and correlation at more detailed levels, however,
rely heavily on the use of assemblage faunas within small areas. Thus, in
Britain, mammalian faunas have been related to the pollen biozones of
interglacials (Stuart 1974. 1976), a practice which relies on the validity
of such subdivisions, and the degree to which they may be recognised as
belonging to a particular interglacial.

In the Lower Thames Valley it is conventional to recognise two interglacials,
Hoxnian and Ipswichian (Chapter 2), which Sutcliffe (1960) differentiates on
the basis of the mammalian assemblages. Hoxnian faunas have abundant horse,
no Hippopotamus or Hyaena, large fallow deer, and a predominance of
Dicerorhinus kirkbergensis (merki) over *D. hemitoechus*. The Ipswichian
contains abundant Hippopotamus, Hyaena and *D. hemitoechus*, but an absence of
D. kirkbergensis, and small fallow deer. But the Ilford Terrace (Fig. 6-6),
between the Upper Floodplain (Ipswichian) and Boyn Hill (Hoxnian) Terraces,
contains a different assemblage, including an early form of mammoth,
Mammuthus trogontherii. From this it is proposed that an additional interglacial occurs (? Ilfordian), and by implication, that the fragmentary pollen
evidence, by which both Upper Floodplain and Ilford Terraces are dated
Ipswichian (Fig. 6-6) is misleading (see also Fig. 6-5). The Stoke Tunnel
Beds, near Ipswich (Turner 1977), and Barling Terrace in Essex (Gruhn *et al*
1974) may also be the same age. An Uranium-series date of 174 ka on bone
from the Brundon interglacial may indicate an even further additional interglacial (Szabo and Collins 1975).

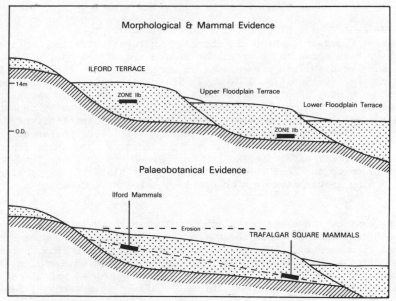

Fig. 6-6.   Interpretation of Thames terraces (Sutcliffe and Bowen 1973)

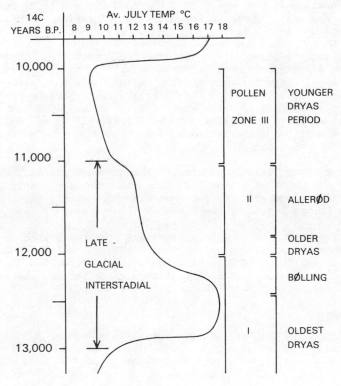

Fig. 6-7.   Late-glacial temperatures from coleoptera (Coope 1977)

COLEOPTERA

The Coleoptera (beetles), which are abundant in fresh-water and continental deposits have been shown to be excellent indicators of climate, and their great mobility renders them pre-eminent in determining the timing of crucial shifts in climate (Coope 1975. 1977). Their robust skeletons allows ready preservation and easy identification at the species level. In morphology they are similar to their modern couterparts, and assumptions of unchanged physiological requirements are amply justified by the consistency of faunal assemblages and the company which they kept. A fastidious preference for specific climatic environments has made it possible to show that migration from high arctic, far east Siberian and southern European areas, characterized Britain during the Last Interglacial and Last Glaciation. Combined with carbon-14 dating they constitute a powerful tool for environmental reconstruction.

29% of the fauna of the Last Interglcial no longer lives in Britain, but occur in southern Europe. This shows that summer temperatures were some 3°C above those of southern England today. Interstadial faunas from Chelford (Brørup) confirm the evidence of the flora in demonstrating a cooler climate than present, with moderate continentality. But between 45 and 25 ka BP, the 'Upton Warren interstadial complex', biotal conflict is apparent , for the beetle data suggest that 43 ka ago summer temperatures warmer than the present day obtained for about 1000 years, whereas trees were absent. The only reasonable explanation is that the sudden amelioration and brief duration of the interstdial was not long enough to allow tree migration. Similar biotal imbalance is apparent in the Devensian Late-glacial. The conventional pollen zonation for the period (Fig. 6-7) identifies the Allerød, Biozone II, as the thermal optimum. Beetle data on the other hand shows that the thermal maximum occurred almost immediately after a sharp rise in temperature ca. 13 ka ago, so that in vegetational terms, it coincided with a herb-tundra environment (Fig. 6-7). About 12,000 BP a decline in temperature occured, and the ensuring cool temperate climate, which lasted for ca. 1000 years, corresponds to the Allerød. Tree migration, therefore, lagged behind thermawarmth to the extent that temperatures were about 3°C lower when they eventually appeared.

The final episode, from 11 to 10 ka ago, the Loch Lomond stadial, witnessed recrudescene of an ice sheet in the Scottish Highlands, and smaller ice-caps and cirque glaciers elsewhere. At the same time high arctic fauns migrated south as far as Cornwall.

The sequence of events for the Late-glacial has been replicated from several sites : south-west Scotland (Bishop and Coope 1977), the Isle of Man, North Wales and Shropshire (Coope 1977). The period of amelioration has been named the Windermere Inderstadial, and its stratotype described and located in the English Lake District (Coope and Pennington 1977).

HOMINIDS

Whereas at one time the skeletal remains of fossil man and his artifacts were used to date Quaternary deposits and events, the reverse is now true. Nevertheless, as the most distinctive and significant fossil lineage of the Quaternary, an outline, however brief, would not be inappropriate. On the one hand investigations have been concerned with taxonomy, the principal evolutionary trends being demonstrated by achievement of bipedalism, smaller canine

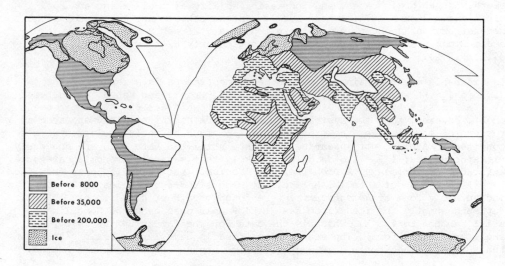

Fig. 6-8.   The spread of Pleistocene man (after Butzer 1977)

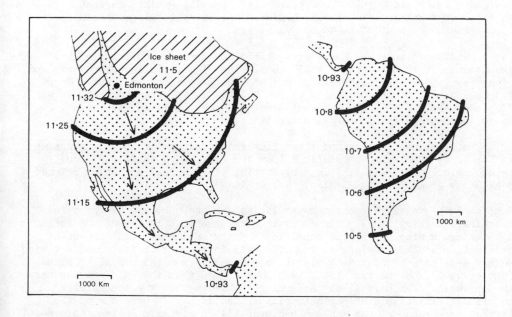

Fig. 6-9.   The population and extinction of the megafauna front in the
Americas.   Dates in 10^3 years BP (after Martin 1973)

teeth, larger brain cases, and increasing ability at tool making and social
organization (Mann 1976. Butzer 1977). On the other hand, palaeoecological
reconstruction has been increasingly important, aided by palaeontology,
sedimentology and radiometric dating, particularly K/Ar methods (Bishop and
Miller 1972).

Considerable doubt remains as to whether *Ramapithecus* is a member of the
Hominidae. Known from India, Pakistan, China, Greece and Kenya, and dated to
14.5 Ma in East Africa by K/Ar, this age is somewhat early. There can be no
doubt, however, that the australopithecines were hominids. They appeared at
the end of the Miocene, about 5.5 Ma, and for much of their evolution were
confined to Africa, not appearing elsewhere until the Late Pliocene and early
Pleistocene, ca. 1.9 - 1.5 Ma. It is likely that several taxonomic categories
existed : *A. africanus, A. robustus,* and *A. boisei,* but they disappear from the
fossil record about 1 to 1.5 Ma ago. At Olduvai Gorge, remains of *A. robustus*
and an early form of *Homo (habilis)* were found together, dated to ca. 1.8 Ma.
But subsequently the finds from East Lake Rudolf push back the Homo lineage
still farther to 2.6 Ma, which is the date for the famous KNM ER-1470 fossil
skull from Koobi Fora. These early Homo fossils have larger brain sizes than
the australopithecines and possess lower limb bones more like modern humans.

Evolution into the taxon *Homo errectus* occurred about 1.5 Ma ago, and fossils
have been discovered in Java, China, Europe, North, East and South Africa.
Brain cases were larger and face and teeth sizes were smaller and closer to
modern humans. During the Middle Pleistocene *Homo errectus* evolved into *Homo
sapiens* - ca. 200,000 to 300,000 years ago. Fossils exhibit further reduction
in face and teeth and show larger brains. Anatomically modern hominids
appeared around about 35 to 40,000 years ago.

The Acheulian culture was the most specific manifestation of Middle Pleistocene
evolution. It involves stone-tools consisting of bifacially worked axes and
cleavers. It was the material culture of hominids living at Olduvai Gorge by
1.5 Ma and much of the Old World by 500 ka BP. In Subsaharan Africa it was
displaced about 200 ka BP, but continued in Europe until as late as ca. 100
ka as the Mousterian of Acheulian tradition.

By 35 ka BP Neanderthalers and their Mousterian culture had been superceded
in Europe by Cro-Magnon populations from the east, with their superior
technological and social skills, culminating in 'Ice Age' art which reached a
peak about 16 ka BP (Magdalenian) (Bulter 1977).

Homo sapiens' vastly increased capabilities are shown by his progress in
colonization (Fig. 6-8). His entry into the Americas by way of the Bering
Land Bridge during the arctic conditions of low sea level, and subsequent
penetration southward, is a matter of debate in respect of its precise timing.
He may well have completed th migration to what is now the U.S.A. by 25 ka
(Butzer 1971), but Martin (1967. 1973) has argued that the evidence for this
is equivocal. Far more likely according to him is that the spectacular
megafaunal extinctions of North America at ca. 11 ka BP were due to the
initial appearance of man who managed overkill on gigantic scales. His model
(Fig. 6-9) is based on considerations of biomass, rates of migration, and
radiocarbon data. His population front swept from Canada to the Gulf of
Mexico in 350 years, and to Tierra del Fuego in ca. 1000 years (Martin 1973).

DEEP SEA BIOSTRATIGRAPHY

The pioneer work of Schott (1935) showed that abundances of *Globorotalia menardii* in oceanic deposits signified climatic amelioration. This was confirmed by the work of Ericson and his colleagues (1964. 1968) who used relative abundances in the *G. menardii* complex to erect faunal zones for the entire Pleistocene, which they correlated with the North American glacial-interglacial sequence. The Plio-Pleistocene boundary was recognised on the basis of extinction of the discoasteridae and, by extrpolation of carbon-14 and uranium series dates, was estimated as 1.5 Ma. This was subsequently revised to 2 Ma, and the boundary redefined to recognise the first appearance of *G. truncatilinoides* in abundance. Their estimate of the age of the Plio-Pleistocene boundary conflicted with that of Emiliani (1955) who had argued that the glacial-interglacial Pleistocene commenced at ca. 400,000 years BP. It was a disagreement pursued at some length. However, while the latest stages of Ericson and Wollin correspond to the oxygen isotope record, earlier ones diverge markedly, particularly below stage W (Imbrie and Kipp 1971).

Faunal extinctions are part of the basis of drawing the Plio-Pleistocene boundary (Chapter 10). In some other instances, however, it is possible to demonstrate extinction levels within the Pleistocene and with reference to oxygen isotope stages. For example, extinction of the coccolith *Pseudoemiliania lacunosa* occurs in stage 13 in both Pacific and Caribbean Oceans.

Lidz (1966) showed that climatic curves based on ratios of a group of species rather than single ones agreed much better with oxygen isotope curves. This approach was quantified by Imbrie and Kipp (1971) who established a series of parameters from foraminiferal assemblages collected from core-tops, thus being assumed to correspond to modern conditions. These included summer and winter temperatures and salinity values. Armed with equations, called transfer functions, they established these parameters for assemblages at different core-depths. This enabled two important trends to be established: (1) downcore variation in climate, (2) geographical movements of ocean surface waters (Fig. 6- 12, 6-13).

Successful applications of this technique have been made in the North Atlantic where thick deposits of undissolved calcareous ooze occur. Periodic influxes of ice-rafted detritus also occur, and the high sedimentation rate allows resolution at intervals of a few thousand years only. Ecologic water masses are those which are defined by a range of temperarure and salinity values and have precise modern analogues (Ruddiman and McIntyre 1976). On this basis Ruddiman and McIntyre showed that eleven major advances of polar water occurred in the past 600,000 years, with ice-rafted debris being recorded at each glacial maxima which may be correlated with independently constructed oxygen isotope curves.

The sole polar species is *Globigerinoides pachyderma* (Fig. 6-11) which has proved particularly sensitive. Using it as an index of climatic change three episodes of rapid change were investigated (Ruddiman *et al* 1977): (1) the cold episode during oxygen isotope stage 7, (2) deglaciation at the stage 6/5 boundary, (3) deglaciation at the 2/1 boundary, which includes the Younger Dryas cold episode. High % of *G. pachyderma* demonstrated low temperatures and salinities, while low % showed high temperatures and salinities. Calibration using carbon-14, interpolation from U-series dates, and identification of a volcanic ash errupted 9300 years BP allowed dating of the core (Fig. 6-11). The data showed that the cold episode in stage 7

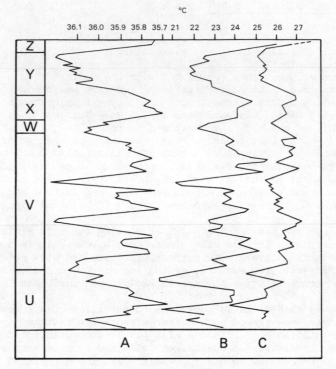

Fig. 6-10. Estimates for average salinity per mil (A), winter temperature (B), and summer temperature (C) of Caribbean surface waters from core V12-122 (Imbrie and Kipp 1971). U-Z : faunal zones of Ericson and Wollin (1968)

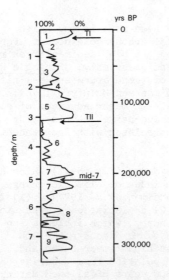

Fig. 6-11. Polar fauna (*G. pachyderma*) as a % of total planktonics. Isotope stages 1-9. T : Termination (after Ruddiman, Sancetta and McIntyre 1977)

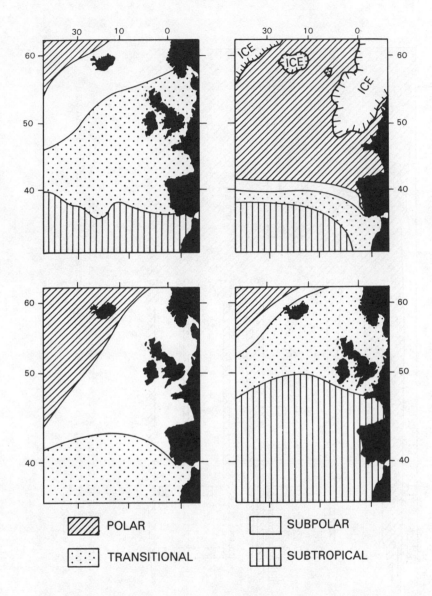

Fig. 6-12.   Maps of the North Atlantic Ocean showing
ecologic water masses at four time levels during the Upper
Pleistocene : Top left : present day. Top right : maximum
extent of the Last Glaciation, ca. 18 ka BP. Bottom left :
ca. 9,3000 BP. Botton right : Last Interglacial 120 ka BP
(after Ruddiman and McIntyre 1976)

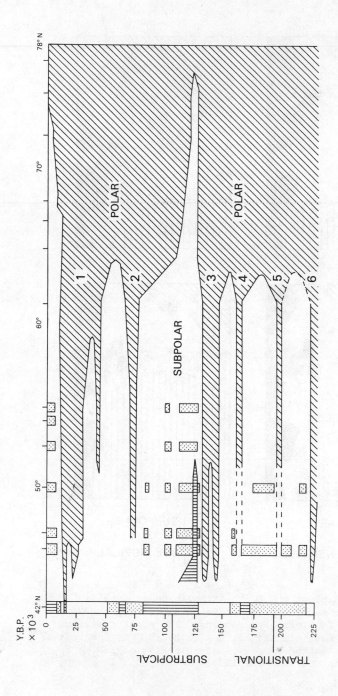

Fig. 6-13. Shifts of ocean surface waters is the North Atlantic for the past 225,000 years. Core stations shown by ticks between latitudes 42 and 78 degrees north (after Ruddiman and McIntyre 1976 and Kellog 1976)

involved a change from interglacial to glacial and back to interglacial
conditions in less then 15,000 years. During that time ice on the continents
built up to 80% of their maximum size and also melted. The glacial-interglacial
warming at the stage 6/5 boundary was even swifter, taking on average 7,000
from one extreme to another, with the first 75% of the change coming in 1700
years, which corresponds to temperature change of the order of 5.2°C/1000 years.
The glacial-interglacial warming at the 2/1 boundary, however, was less abrupt
but the general sequence is remarkable in that it is supported by Coope's
(1977) coleopteran evidence on land. It consists of (1) initial warming at
ca. 13.5 ka BP, (2) cooling after 11.5 ka BP ending about 10.2 ka BP, and
(3) warming mostly before 9.3 ka BP.

MISCELLANEOUS FAUNAS

The remains of Pleistocene animals found in lake and bog deposits were
reviewed at length by Frey (1964). Most, however, are of limited value for
dating purposes, but many useful in palaeoecological reconstruction and for
climatic inference.

The Mollusca, however, are outstanding in that they occur in all environments,
land, freshwater and marine. Bivalves are aquatic, but gastropods occur from
the tropics to arctic and alpine environments. As such they are powerful
climatic indicators and have been used as such in the U.S.A. (Taylor 1965.
Leonard 1953), and in central Europe where they provide valuable corroboration
of soil and loess environments (Lozek 1969) (Chapter 9). In Britain the
interglacials are characterized by similar assemblages of species, but with
some differences. At times of rapid climatic change there is a tendency for
mixed climatic faunas to occur (Kerney 1977).

Marine mollusca, in spite of their ubiquity, have not been exhaustively used
at the same level of investigation like other groups. This arises in part
because their present day ecology is not well know. Some have been used for
local zonation, and others provide information on water depth and climate.
Hydrobia is a brackwish water species that indicates the proximity of marine
conditions, hence may be used to infer marine transgression (e.g. Hollin
1976).

Chapter 7
SEA LEVEL

The interface between continents and oceans, the coastline, is a potential
link between the stratigraphic framework of the deep-sea cores and the often
less than satisfactory sequences on land. A connection between the two has
long been recognised, since Deperet's work on Mediterranean shorelines. Then,
however, it was thought that changes in sea level could be dated by means of
the continental sequence. The reverse should now be the aim of such studies.
The level of the sea is dependent on many variables, and simplistic assumptions,
like those of Zeuner (1959), that sought to establish correlations on
altimetric grounds, have long been abandoned. But the number of instances
when a high degree of confidence may be place in the results as regards dating
and correlation is depressingly few. Those that are, are principally confined to
the Holocene, Last Glaciation and Last Interglacial. Beyond that the number
of uncertainties tends to increase with increasing age.

FACTORS DETERMINING SEA LEVEL

Five principal factors affect sea level and its variation over time : (1) long
term tectonic changes, (2) glacial isostasy, (3) hydro-isostasy, (4)
geoidal changes, (5) glacio-eustatic movements in sea level due to alternating
glaciation and deglaciation.

Menard (1971) presented a map showing the amount of new oceanic crust created
by sea-floor spreading since 10 Ma ago. The late Cenozoic was a time of
continental collision and shortening of coastlines, thus overall regression
occurred. Fairbridge (1961) took this into account when he depicted the trend
of Pleistocene sea level, and Bloom (1971) pointed out that, with ocean basins
widening at the rate of 16 cm a year, since the Last Interglacial the oceans
had accommodated 6% of the returned meltwater despite Holocene shorelines
being some 8 m lower than at 125 ka ago.

The interaction between long term tectonic trends is considered by Flemming
and Roberts (1973) who calculate several models based on plate movements,
crustal subsidence, uplift of mid-oceanic ridges, and increases or decreases
in the lengths of such ridges. Except locally, e.g. in New Guinea (below)
this factor is difficult to evaluate.

Changes of sea level due to glaciation and deglaction have long been
authenticated. Such changes are now subject to independent corroboration of
oxygen isotope curves because these directly reflect changing ice-volumes on
the continents; hence they are a first order indication of glacio-eustatic
changes in the oceans. An implication arising from this is that sea level
during the Last Interglacial, stage 5e, was higher than it had been on any
previous or subsequent occasion (Shackleton and Opdyke 1973). It follows that
raised shoreline features higher than the Last Interglacial strandline must
have been raised to their present positions by tectonic uplift (Shackleton
1975). While this is perfectly reasonable in areas such as California and
New Guinea, it is surprising in the case of Hoxnian marine deposits in England

at ca. 30 m. Posing more problems, however, would be the need to raise world ocean level by that amount.

A sea level model based on Core V28-238 (Chapter 3) compares almost exactly with uranium series dates on emerged coral terraces on Barbados (Mesollela *et al.* 1969). Peaks during stage 5 correspond to coral terraces at 124, 100 and 80 ka BP (Shackleton and Opdyke 1973). Substantial agreement also holds with raised shorelines on Mallorca, and tend to support Butzer's (1975) view that higher ones, above some 50 m, are Matuyama in age. Uplift correction curves (below) fitted to such data, however, are only current approximations, for Stearns (1976) recalculated the Barbados data to show that sea level 124 ka ago differed from sea levels 103 and 82 ka by smaller amounts than those suggested in the original model.

Glacio-isostasy is considered later, but whereas the alternate loading and unloading of the earth's crust by ice has long been appreciated, the effect of loading and unloading ocean floors by water has tended to be ignored since Daly (1925. 1934) first considered it. This has been termed *hydro-isostasy*, and other than on a theoretical basis it is not adequately evaluated. One calculation suggested that broad scale deformation of the earth occurred during the last deglaciation, as water returned to the oceans (Chappell 1974a). It was estimated that during the last 7000 years ocean basins were depressed, on average, by 8 mm, and at the same time continents rose by about 16 m relative to the centre of the earth. More crucial is the conclusion that the effect was infinitely variable for continental shelves of different geometries : for example the model explained the different depths at which older submerged sea levels lay 17,000 years ago -130 to 170 m in Australia and -90 to 130 m off north east America.

The present surface of the oceans is not level, but uneven, for geodetic sea level (the equipotential surface of the geoid) is characterized by undulations of severalmeters amplitude. Past variations of the geoid may account for anomalous sea levels, for example, as during the Holocene transgression (Fairbridge 1961), a matter which has occasioned much discussion.

The geoid is a function of the earth's gravity, which in turn is affected by several other variables, some extra-terrestrial. Maps constructued to depict the geoid show that there is a 180 m sea level difference between the 'high' at New Guinea, and the 'low' at the Maldive Islands (Morner 1976). In a significant review of the relation between past sea levels and possible changes in the geoid Morner (1976) shows how it may have changed with the changes in the glacial and interglacial configurations of the earth. It follows that sea level curves 'of global significance' are of limited value, and if there have been substantial changes in the geoid, as seems likely, then it is possible that high sea levels will have obtained during glacials, and low ones with glacials at given localities. Morner's model already provides an explanation for different Holocene sea level curves, but other factors probably operate also (below).

PLEISTOCENE SEA LEVELS

According to a model of sea level change based on oxygen isotope curves sea level returns to much the same position during successive interglacials. This, however, ignores progressive tectonic effects that are likely with increasing age, although Alt and Brooks (1965) in their study of Florida marine terraces concluded that sea level had never been appreciably above its present height

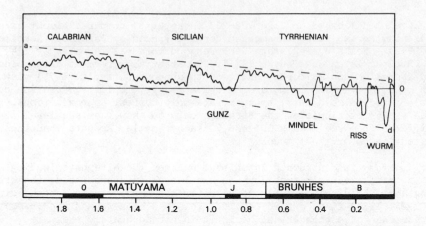

Fig. 7-1.   Sea level change in the Quaternary (after Fairbridge 1971).   O :
Olduvai. J : Jaramillo. B : Blake

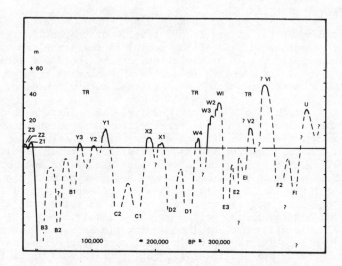

Fig. 7-2.   Relative sea-levels on Mallorca (after Butzer 1975).   Marine
hemicycles : U-Z.   Continental hemicycles : F-B

during the Quaternary. Fairbridge (1961. 1971) outlined a model wherein sea
level descended from Early Pleistocene elevations in a series of glacio-
eustatic oscillations superimpsed on a longer term fall (Fig. 7-1). This was
based primarily on the Mediterranean data, and thus reflected the pioneer
work of de Lamothe and Deperet (Hey 1971). The terminology is based on
biostratigraphic units which hitherto have not been adequately defined in
lithostratigraphic terms. Moreover it is clear that each one includes
several shifts of sea level which reflect changes in climate. Altimetric
correlation is meaningless other than in small areas. Thus Calabrian beds
occur up to 1400 m in south west Calabria, a zone of active plate subduction.
The Sicilian in south west Sicily, however, forms extensive level surfaces
100 to 300 m above sea level, is capped by a regressive beach rock and
significantly, is mostly magnetically reversed (Matuyama). The Milazzian
(later Sicilian) lies about 50 m above sea level in north east Sicily, while
the Tyrrhenian, characterized by the immigrant Senegal fauna, with *Strombus
bubonius*, is capable of subdivision in the more stable regions. Fairbridge
(1973) cites - Tyrrhenian I, 18-25 m (175-230 ka), II, 10-15 m (120-135 ka),
III 2-3 m (65-95 ka). These, however, are merely representative and differ
regionally.

One of the features of Fairbridge's (1971) curve, however, is that low
'glacial' sea levels of the earlier glaciations coincide with the position of
the modern coastline. Thus he argued that low, cold, sea levels, emplaced
exotic erratic blocks on British coastlines and, moreover, suggested that the
wide rock platforms of these and other mid-latitude coasts, were fashioned
under cold climates.

Two outstanding examples of modern approaches to the problems of sea level
change are those located on Mallorca and New Guinea. Interbedded marine and
terrestrial deposits on Mallorca were described by Butzer and Cuerda (1962)
and were shown to be cyclial, consisting of transgressions separated by
regressions when terrestrial deposits accumulated at, and seawards of, the
present coast. These included colluvial deposits (*limon rouges*), terra rossa
derivatives, angular debris and aeolianites which extended from below sea
level to well inland. These regressive beds indicate periods of denudation
and sheetwashing under what were semi-arid conditions.

Subsequently Butzer (1975) departed from the traditional Mediterranean
terminology and instead named 'littoral sedimentary cycles' (Fig. 7-2, Table
7-1). A number of Thorium dates assisted in unravelling a sequence complicated by a multiplicity of sea level stands at similar heights. It was
concluded that this approach was more realistic than one based on altimetry,
using traditional terminology. As the data appeared internally consistent, it
was proposed that indurated beach rock had similar properties to coral in
providing geochemical systems whereby the integrity of the sample is maintained
(See Chapter 5).

The Huon Peninsula of New Guinea is situated near the boundary between the West
Pacific and Australian Plates, and is subject to uplift along a major fault
scarp. Against this lies a spectacular sequence of coastal terraces consisting principally of coral reefs and deltas. The stratigraphic relationship of
the reefs is one of offlap, and the pattern of sea level fluctuations for
each one may be inferred accordingly (Fig. 7-3). This is confirmed by the
palaeoecology of the reef complex which compares with models based on modern
coral associations (Chappell 1974b).

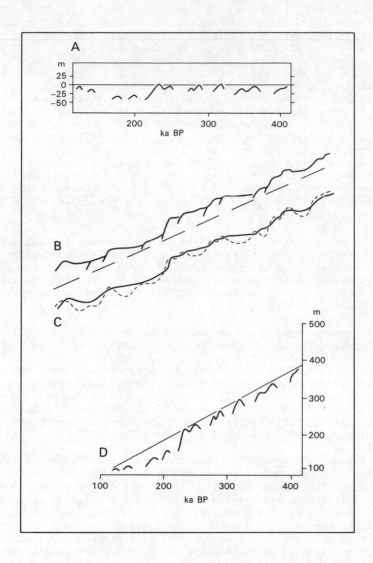

Fig. 7-3.   Emerged coral terraces and reconstructed sea-
levels from the Huon Peninsula, New Guinea (after Chappell
1974).   A : corrected sea-level curve from 100,000 to
400,000 years ago. B : offlap relationship of the terraces.
C : inferred pattern of sea level change. D : graph of
terrace elevation plotted against age and 'uplift correction
curve'

TABLE 7-1   Littoral-sedimentary cycles based on Mallorca
            (after Butzer 1975) (compare with Fig. 7-5)

Marine Cycle	Apparent Sea Level	Fauna	Date ka BP
Z3	2		
Z2	2		
Z1	4		
3 aeolianites		HEMICYCLE B	
Y3	0.5-3		80 + 5
Y2	1.5-2	Partial *Strombus*	110 + 5
Y1	9-15	Partial *Strombus*	125 + 10
2 aeolianites		HEMICYCLE C	
X2	6.5-8.5	Impoverished Senegalese	190 + 10
X1	2-4.5	Full *Strombus*	210 + 10
2 aeolianites		HEMICYCLE D	
W4	4-8		>250
W3	15-18	*Patella*	
W2	22-24	*Patella*	
W1	30-35	?	
3 aeolianites		HEMICYCLE E	
V2 ?	ca.15		
V1 ?	45-50		
2 aeolianites		HEMICYCLE F	
U	30	*Patella*	
	60-65		
	75-80		
	100-105		

The height and age of each reef terrace may be plotted (Fig. 7-3). Age determinations are based on carbon-14 and Thorium dates, samples for both being
subject to stringent field and laboratory scrutiny for recrystallization of
aragonite to calcite (Chapter 5).   Tectonic and eustatic components are
separated by the use of an 'uplift correction curve' (Fig. 7-4), which is
based on three fixed points when the position of past sea level may be
reasonably established.   These are: 120 ka (2-10 m sl), 80 ka (-13 m sl),
and 6.5 ka (-7.5 m sl).   The 120,000 year sea level is known and dated
throughout the world - Barbados, California, Florida, Bahamas, Ryukyus and the
Mediterranean.   After correcting for the difference between such former sea
levels and the present one the uplift correction curve is compared with the
terraces of any given traverse on the peninsula (Fig. 7-3).   Difference in
height between terrace and the uplift correction elevation is a measure of
the height above or below present sea level at which a particular terrace
formed.   After a least squares search, involving several traverses, an
integrated sea level curve is compiled (Fig. 7-3).   This shows sea level
between 100 and 400 ka ago, but Figure 7-5 represents the results of a more
detailed study of the lower terraces, principally of Last Glaciation age
(Bloom *et al* 1974, see also Fig. 10-1).

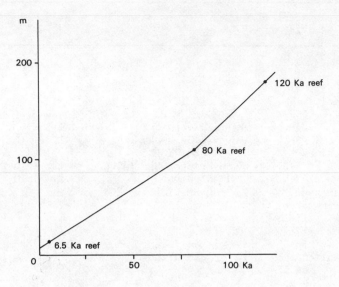

Fig. 7-4.   Uplift correction curve for New Guinea terraces (after Chappell
1974b)

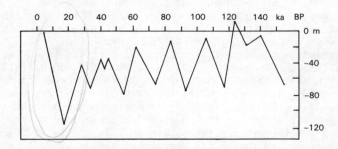

Fig. 7-5.   Sea level change for the past 150,000 years from the Huon
Peninsula, New Guinea (after Bloom *et al* 1974)

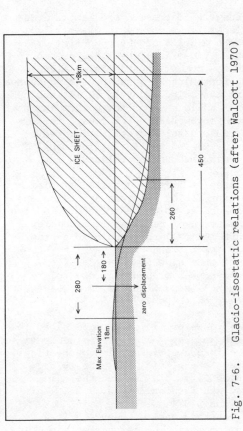

Fig. 7-6.   Glacio-isostatic relations (after Walcott 1970)

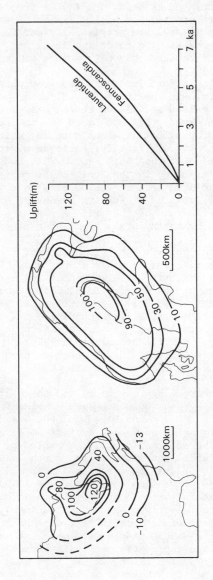

Fig. 7-7.   Glacio-isostatic rebound.   North America 6 Ka BP shoreline (*Littorina*) (centre).   Fennoscandia 7 ka BP shoreline (left).   Postglacial uplift (right) (after Walcott 1973)

LATE GLACIAL AND HOLOCENE SEA LEVELS

The principal effects of a large ice-sheet on the earth's crust are shown in Figure 7-6: the elevation of the ice sheet is 1.8 km, its radius 450 km, and the assumed flexural parameter of the lithosphere 150 km (Walcott 1970). Its profile is based on Weertman (1961), and the maximum stress on the lithosphere is calculated as being ca. 260 km from the ice margin. Three features are noteworthy : (1) the basin-like subsidence beneath the ice sheet, which extends to (2) the proglacial area, for stresses in the lithosphere extend far beyond the limits of the load, and (3) elastic upward bending of the crust (forebulge) beyond the zone of proglacial depression.

On deglaciation, isostatic recovery obtains, the depressed proglacial area recovers, and the forebulge collapses. Isostatic recovery is clearly indicated by the raised and tilted shorelines of North America and Europe (Fig. 7.7). The depressed proglacial area is spectacularly confirmed by the upper shore-lines of proglacial Lake Algonquin, which was not at any time glaciated (Walcott 1970). Finally, collapse of the forebulge is shown by the rapid sea level rise in the north east U.S.A. (where the bulge exceeded 80 m), and by geodetic survey in the Great Lakes region. While much of the uplift inside the glaciated area is due to elastic upward movements in the crust, this is unlikely to account for the forebulge and subcrustal flowage is invoked. These effects, together with a rising sea level and retreating ice-margin, need to be taken into account when considering rebound shorelines.

In comparison with the dimensions of the area infinitely detailed studies have been carried out in central Scotland. Donner (1970) used transgression and regression contacts between peat and marine beds, which were dated by pollen biozones and carbon-14, to elucidate the pattern of relative land/sea-level movements (Fig. 7-8). Up until 8,000 years the falling graph represents the period of elastically rebounding crust. Between 8 and 6,000 years ago the main Holocene trasngression outstripped the then slowing rate of uplift. Finally, uplift continues today, but at slower rates, probably due to relaxation effects in the crust. The sequence is similar, if on a grander scale, for both Fenno-Scandia and North America (Fig. 7-7).

The geomorphology of the raised shorelines, however, is more complicated than Donner's graph would suggest   In the Forth and Tay Valleys the steeply tilted East Fife (1.26) and Main Perth (0.3) shorelines are linked with retreating ice margins which formed during the earlier Late-glacial (Sissons 1976). Following such rapid uplift a period of stability obtained and the Main Late glacial shoreline of the Forth Valley was fashioned (Fig. 7-9). This extensive planation is correlated by Sissions (1974) with the Main Rock Platform of Western Scotland (Gray 1974) that had earlier been ascribed to interglacial or Holocene origin on account of morphological perfection. An interglacial origin was thought unlikely in an area of such intense and deep glacial erosion, and its tilt differs from adjacent Holocene shorelines. Since the Main Late-glacial shoreline was fashioned during the Younger Dryas Period the implications of this correlation are considerable (Chapter 10). The High Buried Beach was emplaced when ice of the Loch Lomond Advance stood at the Mentieth Moraine at 10 ka ago. Subsequently relative sea level fell until the main postglacial transgression, which buried most of the Late-glacial record of the Forth Valley (Fig. 7-9).

Ever since Fairbridge (1961) proposed a Holocene sea level curve with pronounced variations (Fig. 7-8) debate has resounded on its merits by

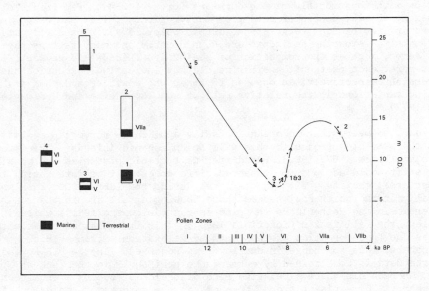

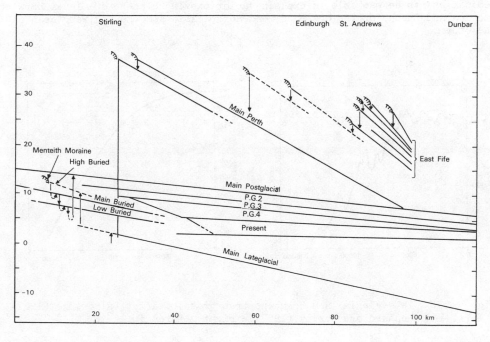

Fig. 7-8.   Relative land/sea level changes in central Scotland (after Donner 1970)

Fig. 7-9.   Raised and buried rebound shorelines in South-west Scotland (after Sissons 1976)

comparison with the smoothly rising curve of Curray (1964). Fairbridge's
curve was synthetic, combining data from many regions, but it has been
confirmed in many studies (e.g. Morner 1969. Tooley 1974). The difficulties
of working with estuarine sediments are considerable and many factors have to
be considered, such as the compaction of peat units (Kidson and Heyworth 1976).
Some interpret each peat unit and marine clay as indicating regression and
transgression respectively, as opposed to those who would regard them as
reflecting variations in the relative rate of sea level rise and sedimentation
(Kidson 1977).

It is clear, however, that no one universally applicable curve can apply.
Perhaps the nearest approach to this will be on oceanic islands - 'Pleistocene
dip-sticks' (Bloom 1967. 1970), but differing tectonic histories and possible
geoidal shift could account for observed differences. It is interesting to
note that areas showing a rise in sea level during the last 6,000 years are
principally in the northern, whereas those showing a fall are in the southern
hemisphere. In the Netherlands the rise is 8 m (Jelgersma 1961), while in
New Zealand the fall is 2 m (Schofield 1960). This is explained by Walcott
(1972) as due to deformation of the earth by relaxation within a thin
asthenosphere, thus no change in sea level is required. In the U.S.A. and
Europe the data is constrained by submergence peripheral to the zone of
rebound, whereas elsewhere a relative rise of the coast occurs. In alternative
view Morner (1976) explains the differences as the outcome of geoidal shift.
A change in geoidal height from high to low in a region would produce
regression and a change from low to high transgression. By using hypothetical
geoidal shifts he was able to explain variations in Holocene sea level between
America and Europe on the one hand, and Australasia and the Mediterranean on
the other.

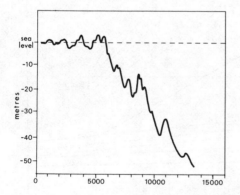 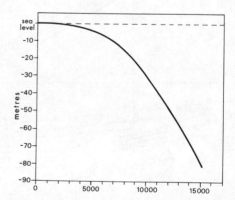

Fig. 7-10.   Late-glacial and Holocene transgression after Fairbridge (1961)
- left, and Shephard and Curray (1967) - right. Dates in 10^3 BP

Chapter 8
GLACIATION

The intrinsic attributes of glaciers and ice sheets, that have figured so prominently in preceding discussions, are now examined. Present day ice sheets, as well as containing a built-in record of climatic change, are useful as depositional models for their Pleistocene counterparts. Such a uniformitarian ideal can never be exact because, for example, there is no present day ice sheet which compares with the enormous Laurentide Ice Sheet of North America. Mid-latitude ice sheets and glaciers, moreover, may have differed in their dynamics, thickness, and thermal regimes, in comparison with present day ones.

Of particular historical significance are the large ice sheets of Greenland and Antarctica, for their compacted layers of ice contain an isotopic record of the Last Glaciation. The isotopic composition of polar ice depends on the temperature of its formation. In Figure 8-1 isotopic values that increase upwards indicate climatic warming, and downwards cooling. The record for two ice sheets is depicted, Camp Century in north west Greenland (Dansgaard *et al* 1969), and Byrd Station, west Antarctic (Epstein *et al* 1970). The isotopic curves are dated using a model of age-depth relationships based on a model of ice flow. Because of the glaciological complexity which obtains at Byrd Station it was necessary to model the relationship according to four different assumptions. The data (Fig. 8-1), however, is broadly comparable for each (Johnsen, Dansgaard, Clausen and Langway 1972).

It is noticeable that all the major climatic fluctuations known from North America and Europe are shown on the curves. The 1390 m long Camp Century core, which reached bedrock underneath the ice, extends back as far as the Last Interglacial, while the Byrd Station one extends to cover most of oxygen isotope stage 5 of the oceanic sequence. Both cores pick up the maximum extent of continental glaciation at ca. 18 ka BP, as well as the cooling of the Younger Dryas period at the end of the Last Glaciation. Fluctuations of the climate in historic time is evidenced in more detailed graphs of the Camp Century data (Dansgaard *et al* 1971), and is verified in comparison with documentary records.

MODELS OF GLACIATION AND DEGLACIATION

Traditional concepts of the rate and mode of ice sheet growth have been summarized by Flint (1971), and may be collectively described as hypotheses of 'highland origin and windward growth', notions still widely accepted. Such a growth involves the initiation and early development of an ice sheet in areas where glaciers occur today, or in high mountainous regions. Glaciation then develops by growth to windward, this reflecting the source of precipit- ation from maritime air masses (Fig. 8-2). In the case of the Laurentide ice sheet the ice divide eventually migrates westwards to the Hudson Bay region, thus reversing the direction of flow so as to inudate and cross the coastal mountains of Labrador. Finally, deglaciation occurs through ice thinning, with surviving remnants being located essentially where the entire process commenced.

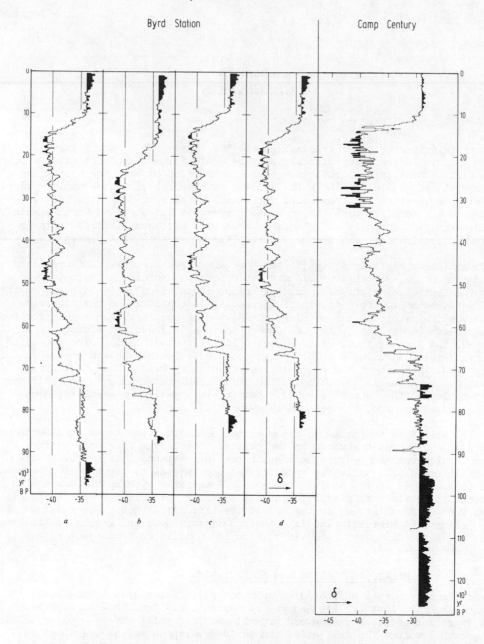

Fig. 8-1.   Continuous isotopic record through the Greenland Ice Sheet at Camp Century, and Byrd Station, Antarctica based on four different assumptions on ice-flow characteristics (modified from Johnsen *et al* 1972)

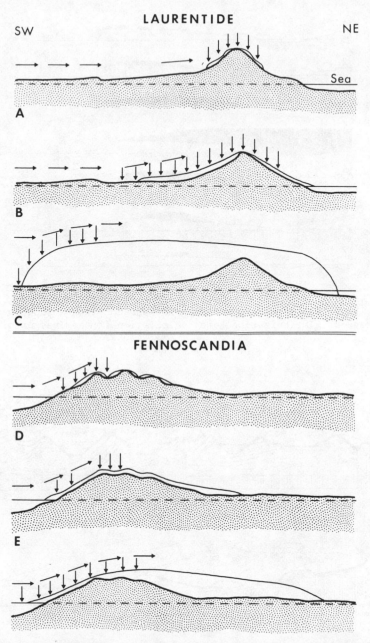

Fig. 8-2. Models of Highland origin and windward growth of the Laurentide and Fennoscandian Ice Sheets (after Flint 1957)

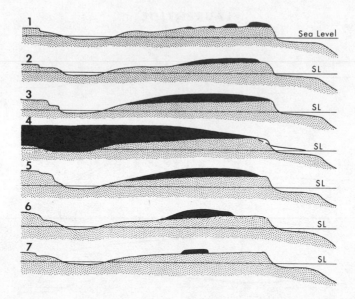

Fig. 8-3. Profile across Hudson's Bay and Labrador-Ungava showing stages
of 'instantaneous glacierization' and deglaciation (after Ives *et al* 1975)

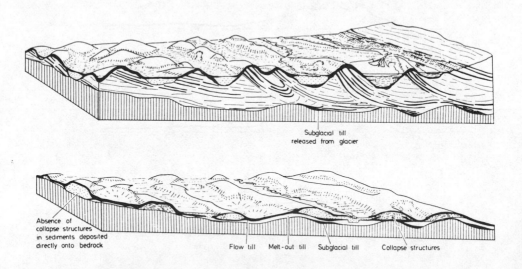

Fig. 8-4. Supraglacial, englacial, subglacial and fluvio-lacustrine
sedimentation (from Boulton 1972)

Data from hitherto unexamined areas in Labrador-Ungava and Baffin Island
(Ives *et al* 1975), as well as the constraints of timing imposed by knowledge
from deep-sea cores, have compelled an overhaul of this model. In alternative
view, a notion of 'instantaneous glacierization', based on the eastern
Canadian Arctic and Subarctic, involves rapid glacier growth initially by a
coalescence of snow banks over wide areas (Fig. 8-3). According to Ives *et al*
(1975) : 'extremely rapid, even catastrophic, events must be envisaged for
both the initial growth and dissipation phases'. When checked against a three
dimensional model of ice flow, and temporal control provided by a recorded
fall in sea level between -20 and -70 m 115,000 years ago in Barbados (Steinen
et al 1973), the model shows that large ice sheets can develop in ca. 10,000
years (Andrews and Mahaffy 1976).

Based on the geologically established extent, and estimates of regime and
temperature, the former profiles, dynamic regime and thermal regime of
Pleistocene ice sheets have been modelled (Weertman 1961). Such exercises are
subject to two principal constraints : firstly the quality of the geological
control for establishing ice extent (below); and secondly, the possibilities
allowed by what is known of the physics of glaciers (Patterson 1969). A model
of the Late Devensian ice sheet in Britain is characterized by a maximum
elevation of 1800 m, and calculations that some 15,000 years would have been
necessary for its growth (Boulton *et al* 1977). This is somewhat longer than
the time calculated for the growth of the considerably larger Laurentide ice
sheet (above).

Modern glaciers as depositional models: Many instances discussed previously
(Chapter 2) show that sedimentary sequences deposited by ice sheets are subject
to erroneous interpretation. At one time any kind of till was invariably
interpreted as a lodgement deposit beneath the sole of a glacier. Till,
however, forms in a variety of ways, and its genesis is exhaustively considered
in the proceedings of two symposia (Goldthwait 1971 and *Boreas* 1977), by
Price (1973), and in a series of papers by Boulton (e.g. 1975). The stratigraphic implication of till and related fluvioglacial sequences is considered
here.

'Count from the top' procedures in particular have reinforced what is an over-
simplified view of glacial-depositional processes. Traditional views
recognize a two fold classification : lodgement till, that is emplaced beneath
the glacier sole, and ablation till, produced from the ablation (melting) of
englacial and supraglacial layers (Flint 1957). Work based on modern glaciers,
however, shows that depositional sequences may be infinitely variable
(Boulton 1972), and more complex classification have emerged (e.g. Francis
1975).

Using modern arctic glaciers, notably from Spitsbergen, as models for
Pleistocene ice sheets, Boulton (1972) showed how contemporary depositional
sequences are analogous, even in detailed respects, to those that were produced
by former ice sheets. He showed that when a glacier containing large
quantities of englacial material decays, supraglacial deposition ensues as
the debris melts-out. This leads to the formation of hummocky till surfaces
and till plains. Boulton's model is shown in Figure 8-4 : ice cored moraines
control the pattern of fluvial and lacustrine deposition, principally within
constricted linear basins, in the decaying ice mass which lies in the distal
zone of the glacier. Flow till, released by melt-out on the ridges, flows
down their sides into the troughs, where it becomes interbedded, or overlies

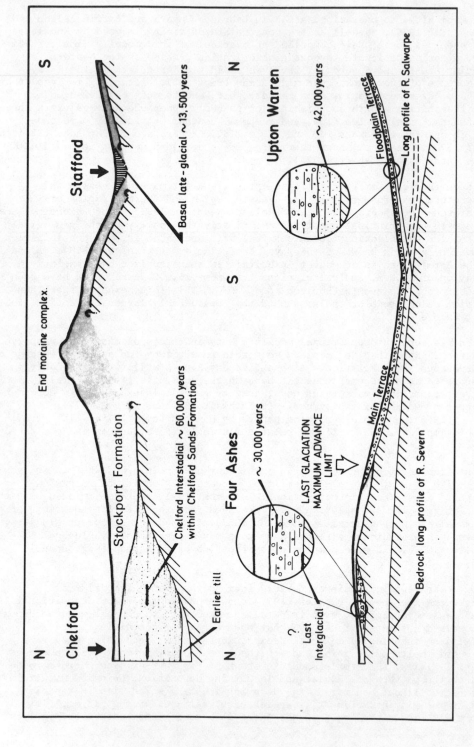

Fig. 8-5. Stratigraphic-geomorphic relations in the Cheshire-Shropshire Plain (from Worsley 1977)

fluvio-lacustrine sediments. Till is also released by melting at the base of
the ice and also within it.

On deglaciation the resultant stratigraphic sequences, especially those of
till-outwash-till (the *tripartite* arrangement) typify those that have been
used to postulate two separate glacial advances, separated by retreat, of
Pleistocene ice sheets. Morphologically, the pre-existing topography has been
inverted, with former fluvio-lacustrine deposits now forming narrow linear
ridges which, in Pleistocene contexts, could be mistaken for end-moraines.

Complexity of depositional sequence, and the formation of ambiguous landforms,
is characteristic around the margins of modern glaciers. However cautious it
may be, the uniformitarian message is clear.

PLEISTOCENE DEPOSITIONAL SEQUENCES AND LANDFORMS

The British Isles: The case of the Cheshire-Shropshire lowland and adjacent
areas (Fig. 8-5), glaciated by an Irish Sea Ice Sheet from the north, and
impinged upon in the west by a Welsh Ice Sheet, is particularly instructive
(Worsley 1976. 1977), particularly as it is the type area for the Devensian
stage. Traditionally (Shotton 1967a) the Last Glaciation was subdivided into
two glacial advances : an early one at ca. 40 ka BP, and a later one after
28 ka BP. This was based on a combination of evidence, especially geomorphic
considerations. Outwash forming the Main Terrace of the River Severn is
apparently coeval with a terrace on one of its tributaries, the Salwarpe where,
at Upton Warren, organic material in the terrace is dated 42 ka BP (Fig. 8-5).
This dated the early Devensian advance. The extent of the later advance was
delimited by the moraine complex that extends from Wrexham to Ellesmere to
Whitchurch and Bar Hill (shown as readvance W on Fig. 8-6). Its late
Devensian age was confirmed by a carbon-14 date of 28,000 BP on a derived
marine fauna collected from glacial deposits north of the moraine. Thus
geomorphic features, terrace and end-moraine, were the primary basis of this
subdivision.

Subsequently the discovery of the Four Ashes Gravel (age Early Devensian to
30 ka BP) beneath Irish Sea till showed that only one glaciation occurred
(Shotton 1967b). This stratigraphic demonstration implies that the Main and
Salwarpe Terraces are not coeval; the carbon-14 date merely indicates early
aggradation at Upton Warren, any correlative in the Severn Valley would
probably have been swept out during the initial discharges of meltwater over-
flowing from proglacial Lake Lapworth by way of the Ironbridge Gorge (Worsley
1977). Sections in the Whitchurch moraine indicate that it may represent a
fluvio-lacustrine depositional system (Fig. 8-4). In fact Boulton (1972)
suggested it may be a supraglacial esker system. Other stratified, ice-contact
landforms, moreover, such as the Salpausselka system in Finland could be of
similar origin. The same applies to the parallel fluvioglacial ridges of
Denmark and North Germany that are currently interpreted as push or overridden
moraines (Chapter 2).

Potentially erroneous reconstructions of former ice limits is a particular
hazard, and is not surprising given the range of inherently misleading
depositional sequences and ambiguous landforms that exist. The extent of the
Last Glaciation has long been a matter of contention. Using 'fresh' landforms
as indicators of its maximum extent, the limit was successively placed at the
tip of the Holderness and Flamborough Head in Eastern England (Fig. 8-6). But
subsequent stratigraphic evaluation showed it had extended to North Norfolk.

Fig. 8-6. Limits of Late
Devensian glaciation after
various workers, and postulated readvances. Eastern
Britain : a, Suggate and West
(1959); b, Valentin (1958);
c, Farrington and Mitchell
(1951). Western Britain : d,
Bowen (1973b); e and broken
line, Charlesworth (1929); f,
Wirtz (1953); Ll, Llŷn (Bowen
1973b); W, Wrexham-Whitchurch
and Welsh Readvance

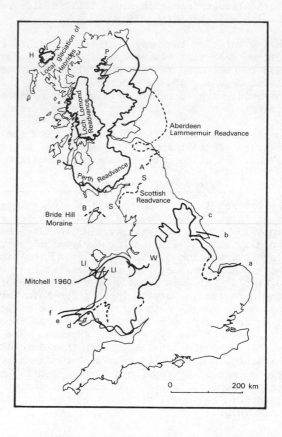

Fig. 8-7. Moraine ridges in Illinois,
U.S.A. (Frye and Willman 1973).
Numbers refer to Fig. 8-8

Similarly in the west, it was successively placed in north Pembrokeshire
(Dyfed), and the Llŷn peninsula, on the basis of 'fresh' landforms, and
periglacial indicators (Chapter 9). Stratigraphic analysis, however, showed
that it had extended farther south. Landform style therefore is an unreliable
indicator and stratigraphic validation should always be sought.

Similarly, several readvance limits have been proposed for the Late Devensian
Ice Sheet (Fig. 8-6). All of them, with the exception of the Loch Lomond
limit of the Younger Dryas, have been invalidated. Some dispute, however,
still obtains on the reality of the Loch Lomond readvance and arises because
unequivocal stratigraphic verification is not yet available (Sissons 1976.
Sugden 1970).

In Lincolnshire a readvance was postulated on the geomorphic position of the
tills (Straw 1969), but textural and mineralogical work showed them to be the
same (Madgett 1975). Farther north the same work showed the Hessle Till of
East Yorkshire to be merely the weathered facies of underlying Drab and Purple
Till units. The Scottish Readvance in Cumbria, and that recognised in
Gwynedd in north west Wales are based on tripartite sequences. That in
Gwynedd was examined by Boulton (1977) who concluded that only one advance of
ice was represented.

So-called end-moraines of the Lammermuir-Stranraer readvance in Scotland were
shown to consist of fluvioglacial landforms in the Lammermuirs (Sissons 1961).
Farther north in Aberdeen a readvance notion based on distinction between
landforms was shown untenable by Clapperton and Sugden (1972) who demonstrated
that the fluvioglacial features of the region were incompatible with such
evidence. The Perth Readvance, prominent in the literature until recently,
was based on sequences of outwash overlying raised marine deposits in that
area. It seems likely, however, that both deposits are contemporaneous, the
gravels having been deposited as deltas built outwards into the Late-glacial
sea (Paterson 1974).

Thus a variable set of criteria used to establish possible readvances of the
Devensian ice margin have been shown to be erroneously conceived. It is
necessary to demonstrate deglaciation - clear signs of ice withdrawal,
preferably on organic evidence, before notions of readvance can be sustained.
Such rigorous criteria are fulfilled in the Great Lakes province of North
America.

The United States: The interpretation of moraine ridges in Illinois (Fig.
8-7) was briefly considered in Chapter 2 (p 50). Willman and Frye (1970)
mapped a series of moraine arcs and identified at least 32 episodes of moraine
construction caused by repeated forward movement of the ice margin. These
were constructed in only ca. 6,000 years, hence glacier advance at the rate
of ca. 0.5 ca. km per year occurred (Fig. 8-8). According to the interpretation the successive moraines are a response to a series of climatic
pulses. On the other hand Wright (1976a. 1976b) argued that they were
produced by a surging ice-sheet, and that evidence of subglacial meltwater, a
lubricant for the ice-sheet base, was not inconsistent with this view.

A new wave of investigations of till lithostratigraphy of the Wedron Formation,
however, often using supraglacial sedimentation models (e.g. Lineback 1975.
Bleur 1974. Hooten 1972), has suggested that a major revision of the Willman
and Frye (1970) model is necessary. Correlation of till units is significantly
different from correlation based on the morainic ridges. Figure 8-8 shows the

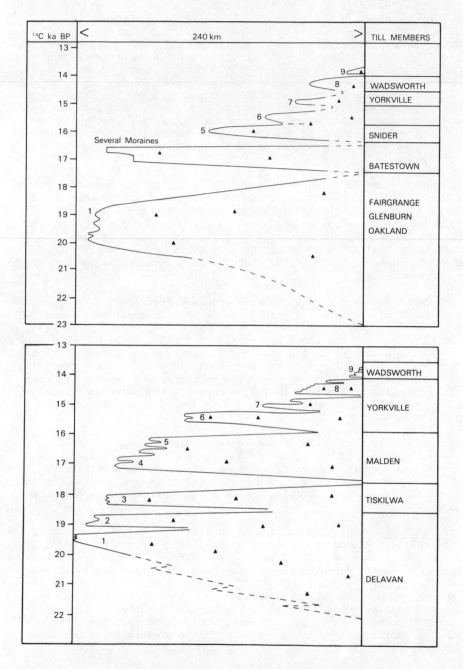

Fig. 8-8.   Two interpretations of the moraines and till units in N.E.
Illinois (Frye and Willman 1973 - bottom. Johnson 1976 - top). Moraines
1-9 (see Fig. 8-7)

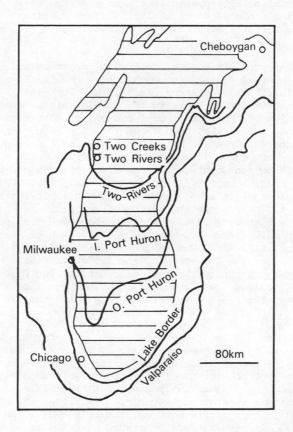

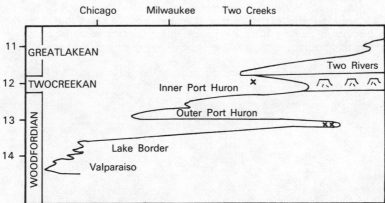

Fig. 8-9.   Ice-margins (top), and time-distance diagram for the Lake
Michigan basin (after Evenson *et al* 1976)

 x : Two Creeks Forest Bed

xx : Cheboygan Bryophyte Bed

major differences in the interpretation of the moraines (Fig. 8-7), the
principal ones being numbered so as to facilitate comparison. It is necessary
to state, however, that the newer hypothesis remains, as yet, untested by
carbon-14 dating.

It is a common experience, therefore, to find major revisions of regional
stratigraphy and inferred Pleistocene events as a result of new discoveries
as well as the acquisition of primary field data. For example, field mapping,
and sub-lake seismic investigations, in the Lake Michigan basin have required
that a major revision of the Late-glacial history of that region be made. As
a result of this new data it has been shown that the major deglaciation took
place not during the Two Creekan but somewhat earlier and amounted to some
400 km in the Cary-Port Huron Interstadial (Fig. 8-9). This is shown by the
Bryophyte Bed of Cheboygan County, which consists of a thin deposit of
mosses and leaves of *Dryas integrifolia*, dated to ca. 13,000 BP (Farrand *et al*
1969). A major readvance followed, extending south to the Port Huron limit
(Fig. 2-22), when ice moved into the waters of glacial Lake Chicago at its
195 m (above s.l.), Glenwood, level. The red Shorewood Till of this advance
is correlated with a submerged end-moraine on the floor of Lake Michigan, as
well as with the red till at Milwaukee (formerly thought to be the post Two
Creekan Valders Till). The Two Creeks deglaciation is considerably less
significant than was previously thought, and when the Two Rivers Till (post
Two Creekan) was deposited the associated readvance was some 175 km less than
that originally proposed for the Valderan Substage (Evenson *et al* 1976).

Because the Valders Till at its type locality is now known certain to be Woodfordian it has been necessary to re-name the post Two-Creekan advance, and the
term Great Lakean has been proposed (Evenson *et al* 1976). Recalling earlier
discussion on glacier margin fluctuations during Late Wisconsin time, this
interpretation does not require major climatic deterioration for the post
Two Creekan advance, a position wholly consistent with climatic inferences
drawn from the regional vegetation as recorded in pollen diagrams (Cushing
1967. Davies 1969). It does not, however, entirely remove the possibility
of glacial surging. Indeed it could be argued that the problems of extensive
deglaciation followed by equally extensive readvance have merely been transferred back in time to the late Woodfordian by recognising that the major
deglaciation occurred during the Cary-Port Huron Interstadial.

Chapter 9
NONGLACIAL ENVIRONMENT

Although the unglaciated regions of the world comprise most of its landsurface, evidence for Quaternary events is, with the exception of quite localised areas, surprisingly sparse in comparison with the glaciated lands. Under
the heading of *nonglacial* comes evidence such as palaeosols and periglacial
features that occurred both within and outside the glaciated areas.

TEPHROCHRONOLOGY

Thorarinson (1954) coined the term *tephra* to refer to pyroclastic materials
transported by air. In New Zealand, however, it has been extended to include
tephra flows and flow breecias from *nuee ardente* eruptions because they grade
into true airfall deposits. As marker horizons tephra layers are probably
without equal because they mark a moment in geological time. Correlation is
effected by carbon-14 dating of associated carbonaceous matter, or more
directly by K/Ar or Fission track dating; and by 'finger-printing' a
particular tephra according to its intrinsic character.

Correlation by such 'finger-printing' includes a variety of techniques : for
example, chemical, mineralogical, optical, mafic mineral content, spectro-
graphic investigation to determine element content, amount of water in the
hydration layer and activation analysis of glass separates. In this way the
widespread extent of several ashes has been confirmed. The Mazama ash of the
Cascades Range has been shown to exist in Oregon, Washington, Alberta and in
Pacific Ocean cores (Borchardt 1971). Field techniques are described by
McCraw (1975) and in some cases allow formal lithostratigraphic definition.
The use of marker beds has been particularly successfully employed in New
Zealand (Vucetich and Pullar 1964). Some difficulties still remain, however,
especially related to the separate identification of tephra of different ages
from the same event.

In the western U.S.A. volcanoes of the Rocky Mountains and Cascade Ranges have
been intensively studied. Thus, in the area west of Yellowstone National Park,
three large calderas and their associated tuffs characterize the celebrated
Pearlette Ash family : Huckleberry Ridge (Pearlette B) ca. 2 Ma; Mesa
Falls (Pearlette S) 1.2 Ma; and Lava Creek (Pearlette O) 0.6 Ma. These enable
dating of the montane glaciations in the region (Richmond 1976) and are shown
on the correlation table (Chapter 10). Elsewhere in the States Izett *et al*
(1972) dated the Guaje Asha at Mt Blanco to 1.4 Ma (fission track date), and
in a review of the Pearlette ashes, showed that Pearlette S was seemingly
magnetically reversed, Pearlette O has normal magnetization and is post
Kansan to early Yarmouthian, and Ivingtonian, in age. Another important
stratigraphic marker is the Bishop Ash, known from several localities between
California and Nebraska, and dated as 0.7 Ma (Izett *et al* 1970).

In New Zealand an earlier set of tephra occur mostly between south Auckland
and Waikato, the earliest being associated with ignimbrite eruptions. Dating

is by magnetostratigraphy from adjacent ocean cores, and fission track dating :
for example, the Mt. Curl tephra gives a date of 230 ka (Milne 1973). A later
set of tephra are dated from 42 ka, the most widespread emanating from the
Muroa centre, Lake Taupo, e.g. The Aokautere Ash, ca. 20 ka, which possibly
carried as far as the northern Canterbury Plains (McCraw 1975).

Tephra erupted in the Rhineland are, unfortunately, not unequivocal indicators
of correlation nor age. Brunnacker (1975) maintains that either the K/Ar
datings are incorrect, or correlation of the eruptions with the terrace stratigraphy is faulty.

Tephrochronology is a critical means of dating and correlating glacial advances
(Richmond 1976), sea level changes, e.g. in Japan (Machida 1975), as well as
dating alluvial fans and rates of coastline progradation (McCraw 1975).

PALAEOSOLS

A *palaeosol* is a soil that formed on a past landsurface (Ruhe 1965). It may
be buried or have remained on the ground surface (relict). Because the term
has been loosely used much unnecessary confusion has arisen. It has, for
example, been used for soil *in situ*, as well as redeposited soil (colluvium)
within which the soil fabric is frequently preserved. It has also been applied
to both the weathering and soil profiles. This is incorrect because a soil is
the end-product of weathering under the influence of climate, biota, topography and time, and has horizons differentiated by the relative intensity of
chemical processes and the translocation of products, quite unlike the results
of rock weathering. Moreover, the latter often occurs at the interface
between an impermeable and permeable layer and does not signify subaerial
exposure.

Burial and diagenesis may alter a soil considerably. The fact is that soils
possess few unique features as opposed to sediments : even clay skins are
common both (Valentine and Dalrymple 1976). Thus even palaeosol identification
is not without problems.

Palaeosols are used for environmental reconstruction (see Loess) as well as
for correlation and dating. Correlation should depend on full knowledge of
palaeosol variation geographically, because without this it is likely that
genetically similar soils at close but different stratigraphic levels will
be miscorrelated. Thus it has been suggested that instead of a type site for
a soil stratigraphic unit there should be a type transect (Brewer *et al* 1970).
In practice, however, most correlations assume that the effects of universal
and synchronous climatic variations will have broadly comparable effects on
soil development in widely separated regions. For example, Morrison and
Frye (1965) correlated the Cocoon Soil of the Great Basin, Nevada, with the
Sangamon Soil of the Middle West, largely on account of its strong development
in both regions.

Recognition of palaeosols in loess, terrace and pluvial sequences of former
lakes, is of critical importance in subdividing the record. A noteworthy
example of the use of palaeosols in subdividing a lengthy glacial sequence is
that in the La Sal Mountains (Richmond 1962). This is one of the few studies
to evaluate the terrain effect on lateral and altitudinal variations in soil
type. The area rises above the Colorado Plateau (2100 m) to 4170 m, with
accompanying vegetational variation from sage brush grass to scrub oak to
montane aspen to spruce-fir to Alpine, and soils from Sierozems at lower

elevations through Brown Forest soils and Brown Soils to Brown Podzolics at
the higher elevations. The practice is to rank the soils in a hierarchy
according to their relative degree of profile development : that is, into
categories of weak, moderate, and strong. Then the hierarchy is matched with
a similar one in the region with which it is intended to effect correlation.
Usually the basic framework is underpined by the strongest soil, most often
the Sangamon.

Soil stratigraphic correlations in the western United States are summarized by
Birkeland (1974) who comments on some of the difficulties. As with other
phenomena that tend to be repeated time and again in the Quaternary the
dangers of homotaxial error loom large. In retrospect it would seem that the
earlier use of palaeosols for subdividing the classical sequence of central
North America outran the state of soil science at the time. Currently
increasing knowledge of present soil forming processes serves to emphasize the
inadequate basis of many such early correlations.

PERIGLACIAL ENVIRONMENTS

The term periglacial refers to environments where frost action is the dominant
process. Such environments occur at high latitudes and high altitudes today
both inside and outside the zone of permafrost. For European mid-latitude
areas it is clear that a periglacial climate was the norm for most of Pleistocene time, hence the importance of its associated processes*. The range of
these includes slope, alluvial, aeolian as well as a variety of structures due
to the growth of ground ice. While the present day permafrost environment
provides some analogues for Pleistocene ones, it is often less than adequate
for mid-latitude areas. There is the angle of the sun during the Pleistocene
would never have been so low as it is farther north today. This had an
important effect on the rate of operation of periglacial processes. The fact
remains then, that no precise contemporary periglacial analogue exists for
Pleistocene mid-latitude regions.

Slope deposits (Benedict 1976) are strongly influenced by factors of site,
including bedrock lithology and availability of superficial deposits for
redistribution downslope.   Often, however, it is difficult to distinguish
periglacial slope deposits from till. In cases of doubt the term *diamicton*
should be used. Potential confusion is not surprising : both are unsorted,
both may be stratified, and in some cases older glacial material has been
incorporated into that of periglacial origin. By careful fabric and grain
size analysis periglacial diamicton was differentiated from the Chalky Boulder
Clay till of Essex by Baker (1976).

Ambiguity of origin posed by certain diamicons has fundamental bearing on the
sequence of Pleistocene events in certain regions. Diamicton interpreted as
till overlies certain interglacial beaches in south-west Britain and points to
the last local glaciation of such areas as having occurred subsequent to the
high sea level event. Yet alternative interpretation demonstrates the deposits
to consist of redistributed old glacial material and shows that the last local
glaciation antedated the raised beach event (Bowen 1969. 1973). Using
diamicton, interpreted exclusively as periglacial in origin, the Isle of Man
(Thomas 1976), and central Wales (Watson 1977), have been identified as having

---

* The full range of periglacial processes, deposits and structures is discussed
  by Washburn (1973) and French (1976).

been unglaciated during the Late Devensian glaciation. This led to highly
improbable reconstructions of the Late Devensian ice margin, and flew in the
fact of established stratigraphic and carbon-14 evidence. The alternative
view interprets the diamictons as essentially, though not exclusively,
composed of glacial material, redeposited during and after deglaciation,
particularly during Younger Dryas time (Potts 1971. Bowen 1973a. 1973b. 1974),
thus highlighting the extreme rapidity of periglacial processes.

Ice wedge casts and pingo scars indicate former permafrost - no more and no
less. Yet they have been misused as stratigraphic artifacts in delimiting
ice-margins. Ice wedge casts, it was proposed, only occurred outside the
limit of a readvance in Scotland (Peacock *et al* 1968), whereas they are known
from well inside that limit (Sissons 1976). Similar criteria, together with
the occurrence of pingo scars, was used as a means of recognising unglaciated
areas during the Late Devensian in Wales (Watson 1972), despite pollen
analytical and carbon-14 data that showed that pingo growth occurred during the
Younger Dryas (Handa and Moore 1976).

These matters have been considered at some length because the implication of
process rapidity is considerable. Carbon-14 dating has now established this.
Many spectacular effects have been demonstrated: for example, during Younger
Dryas time alone, large dry valleys in the Chalk scarp were fashioned by niveo-
fluvial processes (Kerney *et al* 1965); incision to 7 m followed by aggradation
of gravels to the same depth occurred at Sproughton, Essex (Wymer *et al* 1975);
alluvial fan deposition obtained in the Longmynd of Shropshire (Osborne 1972);
and extensive solifluction of drumlin slopes took place near Glasgow (Jardine
et al 1976). All these, and others, besides, indicate conditions of rapid
melting and areal denudation of unvegetated slopes.

Periglacial phenomena then, while valuable indicators of environment, require
careful appraisal. Ambiguous sediments can yield erroneous conclusion, and
rates of process operation do not appear to be matched by any present day
environment.

LOESS*

Loess is a calcareous wind blown silt which attains considerable thicknesses
in the former periglacial zone. It is thickest in China and the Ukraine
(Velichko 1969) but is both extensive and thick in the United States and
Europe. It refers to any fine-textured deposit of aeolian origin other than
dune sand or tephra in New Zealand, where its source appears to have been by
deflation from gravel flood plains and an exposed continental shelf (McCraw
1975). But elsewhere deflation from bare unvegetated outwash plains and
gravel aggradations along major valleys would appear to be its principal
source. This is confirmed by its contained gastropod faunas (e.g. Lozek 1969);
such palaeontological data is useful because no sound environmental analogues
for loess accumulation exist today.

Continuous sequences account for its high stratigraphic value in some regions.
In others, however, it is restricted stratigraphically : in Britain it is
mostly early Late Devensian in age, and in the south has often been mixed by

---

* The 'cover sand' lithofacies is discussed by van der Hammen (1951) for the
  Netherlands, and Catt (1977) for Britain.

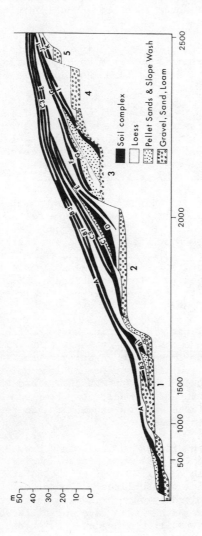

Fig. 9-1. Cerveny Kopec (CK - Red Hill), Brno, Czechoslovakia (simplified from Kukla 1975). Distances in m from R. Svratka. A to K baselines of glacial cycles. Terraces 1 to 5

cryoturbation with other deposits (Catt 1977). Only pockets of early loess
occur, mostly in south east England. Geochronometric dating in New Zealand
shows that loess was accumulating when the Mt. Curl tephra (230 ka) erupted,
while during the later Pleistocene other tephra allow calibration. Carbon-14
of these shows that no loess accumulated in the Banks Peninsula between 30 and
12 ka ago (Runge *et al* 1973), but considerable quantities were deposited
between 12 and 10 ka ago during glacial retreat. In the United States loess
units are well known in relation to till and palaeosol units (Chapter 2)
(Ruhe 1965). But it is in central Europe, in Czechoslovakia and Austria, and
adjacent regions, that the stratigraphic value of loess attains its greatest
significance. Apparently continuous sequences occur; in some instances back
to the Olduvai event.

The Loess sequence of central Europe: By combined data from loess profiles,
interbedded palaeosols, and snail faunas from both, and using carbon-14
dating and magnetostratigraphy, a truly remarkable picture of Pleistocene
events has emerged (Kukla 1970. 1975).

Two other sedimentary units, both of palaeoclimatic significance, are associated with the loess : pellet sands, and 'markers'. Pellet sands are sand-
sized pellets of silt or clay which form during episodes of torrential rainfall following a warm dry season. In many cases they were contemporaneous
with the aggradation of fluvial terrace gravels. 'Markers' are thin light
coloured silt layers, the product of great dust storms.

Soil types, strongly indicative of their contemporary vegetational environment,
are of two main types : biologically reworked prairie soils, with an accumulation of organic matter; and leached Braunerde type soils. In the loess two
cimatic soils occur : Chernozem (confined to the loess belt) and Parabraunerde,
leached and decalified braunerde soils with clay accumulation in the B horizon
(southern parts of the loess belt). Full-glacial soils are the frost gleys
and pseudogleys, which developed on a frozen substrate. Only Braunlehms
(Kubiena 1956), that have Mediterranean affinities, and frost gleys, do not
occur in the Holocene record.

Loess and its associated sediments are rich in gastropod faunas. The climatic
significance of these has been established by Lozek (1964. 1972) who recognised
three principal groups. In the loess, the *Pupilla* fauna, impoverished
numerically and species-wise, indicates cold conditions but the *Columella*
fauna is even colder and includes *Vertigo pseudosubstriata* which today inhabits
the Alpine zone of the Tien Shan under conditions of extreme cold and
continentality. The *Striata* fauna, on the other hand, is representative of
the central European environment today.

A repetitive association of depositional units around Prague and Brno allowed
Kukla (1961) to define 1st and 2nd order sedimentary cycles; and later formal
lithostratigraphic units (Kukla 1970). 1st order cycles are synonomous with
glacial ones, and are delimited by 'marklines'. 2nd order cycles, synonomous
with stadial cycles, are delimited by submarklines. Each cycle and subcycle,
consisting of phases, commences with a thin layer of hillwash sediment : (1)
thin hillwash loam, (2) forest soil, braunerde, (3) chernozem, (4) marker,
(5) pellet sand, (6) loess. Some of these units (phases) may be missing at
given localities, but detailed representative sequences of the last three
glacial cycles occur. The stratigraphic framework is provided by marklines,
boundaries between a thick loess and the overlying parabraunerde or braunerde
soil. Each unit or glacial cycle is designated by capital letters : Cycle A

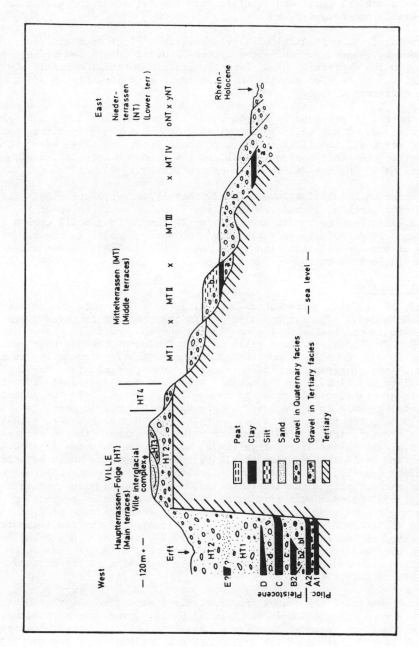

Fig. 9-2.   The Rhine Terraces near Cologne (from Brunnacker 1977)

consists of the Holocene thus far, B, the last Pleistocene cycle etc. Hence
the thermal optimum occurs at the beginning of each cycle, and the coldest
part immediately before its end. Submarklines are boundaries between a loess
and a braunerde, parabraunerde or chernozem. A unit delimited by submarklines
is called a stadial cycle and labelled with the letter of the glacial cycle
and an arabic numeral, e.g. B3.

This practice, while pragmatically valuable, as a first approximation, is not
unattended by danger, for while the cycles hitherto identified differ systematically in the expression of different phases, homotaxial error is always a
possibility. It is, for example, noticeable that no stratotypes are proposed
for the different cycles.

Dating control is provided by carbon-14 on charcoal and soil humus samples as
far back as ca. 30 ka, while magnetostratigraphy has recognised the Brunhes-
Matuyama reversal, the Olduvai Event, and near Brno, the Blake event (Chapter
5). Further than this control, marklines are used for correlation with the
terminations of the oceanic isotopic record (Kukla 1970).

Of the numerous localities investigated, that at Cerveny Kopec (Red Hill)
Brno, Czechoslovakia, is outstanding (Fig. 9-1). Five terraces of the
Svrataka River bear sediments and soils of 10 completed glacial cycles (Kukla
1975), and the stratigraphic relations of the various units to the alluvial
gravels allows age relationships to be established. The Brunhes-Matuyama
reversal is located on top of Terrace CK4.

Alluvial Terraces

The study of alluvial terraces has been hindered in the past due to over-
simplified models. Thus at one time it was the practice to refer given
terraces to former sea levels (Zeuner 1959), whereas now it is know that
terrace height above the flood plain neither indicates age nor past sea level
(Clayton 1977). Equally a model based on warm climatic downcutting as opposed
to cold climatic aggradation is inaccurate. Gravel aggradation takes place
in a variety of climatic environments. A further complication is that terrace
sediments are particularly prone to re-working. Increasingly the granulometry
of alluvial gravels is being used to infer palaeohydrologic conditions (e.g.
Schumm 1965. Baker 1974 - alluvium). Stratigraphically, however, alluvial
terraces are important as indicators of climatic change. In the central
interior of the United States Frye (1973) has compiled a model of alluviation,
incision and periods of surface stability from the stratigraphic relationships
of alluvial gravels, till sheets and palaeosols. He showed that erosion tended
to occur early during a glacial pulse, and deposition, partly by outwash, late
in the glacial pulse. Soil forming episodes were generally periods of
stability although some aggradation did occur. Much the same sort of sequence
occurred in the Avon valley of England. Previously the terrace sequence had
been related to sea level changes (Donovan 1962), with Avon No. 3 and 2
Terraces ascribed to the Ipswichian. Now, however, although some interglacial
aggradation occurred, they are dated as Early Devensian (Shotton 1968) and do
not relate to sea level.

Alluvial deposits of the Rhine: particularly instructive are the Rhine
terraces around Cologne and in the Neuwied Basin, downstream of Koblenz.
Upstream the Rhine terraces have been subject to uplift, whereas downstream
they have been depressed in the southern North Sea region of downwarping.
Around the hinge zone, however, the gravel units are superimposed (Fig. 9-2).

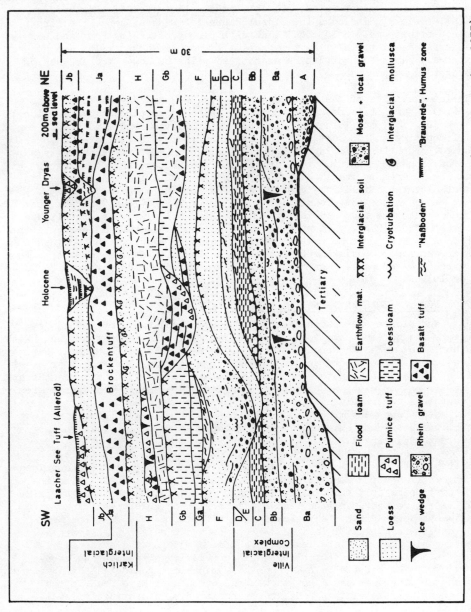

Fig. 9-3. The Pleistocene succession of Karlich and the Neuwied Basin (from Brunnacker 1977)

Traditionally the Upper Middle and Middle Middle Terrace have been ascribed to
the Elster Glaciation, the Lower Middle Terrace to the Saale, and the Low
Terrace to the Weichselian. But after studies of the alluvial sediments,
particularly their quartz ratios, heavy minerals, mollusca, magnetostratigraphy,
loess overs and tephrochronology, Brunnacker (1975) showed that the sequence
was more complicated (Fig. 9-2). The correlation table in Chapter 10 shows
the regional stratigraphic position of these. In passing the following points
may be noted : (1) Main Terrace, HT1 is magnetically reversed (Fig. 9-2),
(2) Main Terrace HT3 includes the oldest known ice-wedges, hence the first
appearance of extremely cold conditions with permafrost, (3) Four and not two
Middle Terraces exist; MT 11a is magnetically normal and contains ice-wedges,
(4) extremely rapid downcutting is associated with the Lower Terrace during
the Younger Dryas period.

The Karlich clay pits of the Neuwied Basin (Fig. 9-3 and Correlation Table in
Chapter 10): here it is possible to relate episodes of terrace aggradation,
loess accumulation, and volcanic activity. Most of the beds consist of
alluvial deposits and loesses. Dating and correlation is indicated in
Chapter 10. But the following key horizons occur in what is essentially a
continuous sequence. Units A to C are fluvial sediments, but from unit D
upwards are covered by loess. The lower part of the 'Cromerian complex' is
represented by the 'Ville interglacial complex' of units Ba to D. The
earliest recorded ice wedge casts in Europe occur in the lower gravels unit Ba.
The Brunhes-Matuyama boundary is located in the Moselle gravels of unit Bb
(Brunnacker *et al* 1976). Volcanic activity is indicated by Acmite-augite in
unit Ba, while from unit Gb upwards brown hornblende occurs. The molluscan
faunas are entirely consistent with sedimentary data. Dates interpolated from
K/Ar, carbon-14 and palaeomagnetic horizons are shown in the Correlation Table
(Chapter 10).

LOW LATITUDE ENVIRONMENTS AND PLUVIAL LAKES

The Early and Middle Pleistocene record of low latitudes is, by and large,
fragmentary and not amenable to large scale analysis and reconstruction. With
the exception of local sequences, often fortunately calibrated by K/Ar dating
and palaeomagnetic horizons, as at Olduvai Gorge and the Lake Rudolf area,
data is sparse and inadequate. No Middle Pleistocene continuous sequences are
known in Africa for example, no cores of organic data exist, and faunas even
today contain a remarkably high proportion of Pliocene elements (Butzer 1975).
However, assemblages are mixed, and few evolutionary lineages known, so that
palaeoecological reconstruction, largely through sedimentology, is the primary
goal (Butzer 1975. 1977).

A similar paucity of data characterizes South America, with local exceptions
as in Bogota (van der Hammen 1971) and Patagonia (Mercer 1976). Even the
Last Glaciation is only known in rudimentary form (e.g. Damuth and Fairbridge
1968). Elsewhere, however, the Last Glaciation is tolerably well known in
Australia and North Africa (e.g. Rognon and Williams 1977). But the best
records derive from pluvial lake basis that recorded shifts in climate by
alternating high and low lake levels. In Africa much of the evidence,
however, is ambiguous, and inter-regional correlation often contradictory
(Butzer 1976). Butzer (1975) has shown how precarious ecological balance and
climatic stability is today : how much more so would it have been in the
rapidly changing late Pleistocene, and overprinted with distinctly local
effects?

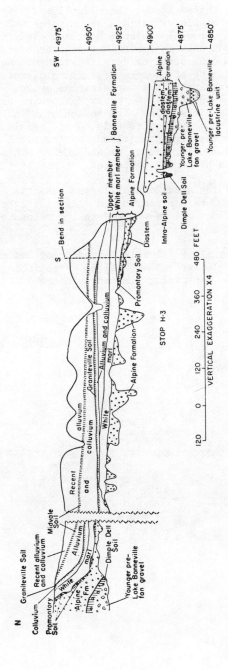

Fig. 9-4. Pleistocene succession of Lake Bonneville at Little Valley, Promontory Proint. Utah (from Morrison 1966)

Pluvial lakes

Pluvial lakes existed in Asia, Australia and Africa, but those most intensively
studied are in the western United States (Morrison 1964. 1965). Modern work
on pluvial lakes utilizes detailed stratigraphic studies and comprehensive
mapping so as to provide a three-dimensional picture. This is necessary
because studies based principally on landforms, such as terraces, deltas and
bars, have not proved good indicators of detailed change. Morrison (1966)
drew an analogy between a stack of saucers of varying diameter, height and
thickness, each one representing a lake depositional cycle, and the stratigraphic arrangement of lake deposits. He evolved a system of lithostratigraphic classification from his studies of Lake Lahontan and Lake Bonneville,
recognising lacustral and subaerial units (Fig. 9-4). Maximum stands of lake
level were determined from the highest altitude of lacustral deposits, and
minima from unconformities, soils, colluvium and loess. This approach revealed
fluctuations of lake levels hitherto undetected.

Older Pleistocene deposits in Lakes Bonneville and Lahontan are deeply buried,
but in the former there is data for Kansan and Illinoian lake cycles. But the
evidence for Wisconsinan fluctuations allows detailed reconstruction (Fig. 9-4,
Correlation Table, Chapter 10).

Such pluvial lake sequences allow valuable extension of sequences recognised
in regions of continental glaciation, as well as establishing stratigraphic
outposts in areas otherwise containing limited evidence.

Chapter 10
OVERVIEW

There is no doubt whatsoever that Quaternary systematics have been subject to
a revolution comparable to that of plate tectonic theory on geology as a whole.
Instead of the Pleistocene comprising four, five, or at most six major
glacials, it is *fact* that eight glacials and eight interglacials occurred
during Brunhes time alone, while altogether some seventeen glacials represent
the entire Quaternary. Existing models of classification are no longer
adequate, and their terminology at best subject to redefinition, at worst
persistent and downright misleading. Existing continental data needs re-
evaluation. For example, Kukla (1975. 1977) has shown that the four classical
stages of the Alpine region span ca. 0.8 Ma, and moreover, they are not due to
climatic cycles but to phases of accelerated crustal uplift.

Encompassing some seventeen glacial cycles in the Quaternary (ca. 1.6 Ma) has
the effect of speeding up the rate at which different processes operated.
Signs of this were clear enough for all to read when the rate of Holocene sea
level rise was calibrated by carbon-14 dating (Godwin and Willis 1958). Since
then it has been shown that large ice sheets may have developed in ca. 10,000
years (Andrews and Mahaffy 1976), which is fully consistent with the data from
the deep sea cores. Other rapid events include the spectacular periglacial
modification of upland Britain during the Younger Drys period (Chapter 9), and
the large scale shifts of North Atlantic surface waters (Ruddiman and McIntyre
1976), paralleled by the remarkable shift in the coleopteran fauna (Coope 1977).
Equally swift changes in biota are revealed by the sequence from deep sedimentary basins as in : Grand Pile, France (Woillard 1975), the Tenaghi-Phillipon
peat bog in Macedonia (van der Hammen *et al* 1972), Lake Biwa in Japan (Horrie
1976), the Sabana de Bogota in Colombia (van der Hammen *et al* 1964), and from
Queensland, Australia (Kershaw 1974).

A re-evaluation has commenced in most disciplines. In archaeology it is now
realised, despite long resistance, that dating and classification by means of
technological typology, for example by stone tools, is no longer possible in
most cases. The Acheulian stone industries of Africa are possibly as old as
1.4 Ma, but in Atlantic and Mediterranean Europe they are rare until mid-way
through the Brunhes Epoch (Isaac 1975). Traditional biogeographical concepts
are being re-assessed (Chapter 6). By analogy with Holocene vegetational
development in North America it is suggested that interglacial patterns never
achieved equilibrium, largely because of the brevity of those intervals
(Davis 1976). This means that substantial regional differentiation of the
flora occurred, thus further devaluing correlation based on assemblage biozones. For example, whereas formerly the Gortian Interglacial of Ireland was
correlated with the Hoxnian of eastern England (Mitchell 1972), it is now
dated, on stratigraphic grounds as Last Interglacial (Warren i.p.)

Instances of rapid, even catastrophic, geomorphic evolution, have long been
appreciated : for example, the draining of Lake Missoula and the fashioning
of the channelled scablands of Washington (Bretz 1956). But rather more

subtle is the evidence for repeated and comparatively brief intervals of
valley incision in central North America, as revealed by the stratigraphic
record (Frye 1973). Along the coast it is suggested that shore-platforms and
cliffs in western Scotland were fashioned in less than 1,000 years during the
Younger Dryas period (Sissions 1974). Similarly, Fairbridge (1971) suggested
that shore-platforms of mid-latitudes were fashioned not during interglacials
but when earlier Pleistocene sea levels were low and sea-ice facilitated
rapid formation.

Episodes of exceptionally high activity have been related in many cases to
times when major shifts in climate were taking place. The suddenness of such
changes is recorded in the Camp Century ice core, where a change from fully
temperate to fully glacial climate occurred in less than 100 years some 90 ka
BP (Dansgaard *et al* 1971), and in a deep-sea core in the Caribbean at the same
time (Kennet and Huddleston 1976). The cause of such rapid changes in climate
is outside the scope of this book, but it may be noted that of the tens of
theories available in the 1950's, a large majority have fallen by the wayside
(Flint 1971). Current ones need to take into account not only the repetitive
character of Pleistocene climates, but also the suddenness in the change from
one climatic state to another : e.g. Hollin's (1965. 1976) surges of the
Antarctic ice sheet triggering off global cooling, meteorologically modelled by
Flohn (1974). But the apparent confirmation of the basic Milankovitch
mechanism by the CLIMAP group is a major achievement (Hays, Imbrie and
Shackleton 1976). With the reconstruction of the earth's surface at various
times in the Pleistocene, e.g. at 18 ka BP (CLIMAP 1976), the stage is set for
meteorological modelling (Williams 1975).

STANDARD CHRONOSTRATIGRAPHIC SCALE

Because all continental sequences contain actual and potential hiatuses the
deep sea cores must be used as standard. All are available for inspection at
various institutions, those of V28-238, V16-205 (Chapter 3) and K708-7 are at
the Lamont-Doherty Geological Observatory, New York. It may eventually be
possible to designate stratotypes on land, but that time has not yet come.
Core V28-238 covers all the Brunhes Epoch and part of the Matuyama Epoch. It
has been proposed as standard for the later Pleistocene (Shackleton and Opdyke
1973), and is used as the basic stratigraphic framework for Table 10-1.
Although core V16-205 extends to the base of the Pleistocene and beyond (van
Donk 1976), no such proposals have yet been made for the earlier Pleistocene.

Definition of the base of the Quaternary has long, and still continues to
arouse debate. At the 1948 International Geological Congress in London it was
recommended that it be defined at the base of the marine Calabrian Stage in
Italy, the stratotype for which is located as Santa Maria di Catanzaro,
Calabria. Several criteria for its recognition were proposed, including the
appearance of *Globorotalia truncatilinoides*, but a recently revised estimate
showed this to be a 'weak reed'. Instead, based on calcareous plankton *datum-
events* and 'multiple overlapping criteria' it is now placed at a level slightly
younger than the top of the Olduvai Event (Haq, Berggren and van Couvering
1977). But regardless of the detailed micropalaeontological argument, the
base of the Pleistocene in Table 10-1 is drawn at the top of the Olduvai
Event, a boundary of world-wide significance, and hence a desirable one for
purely pragmatic reasons.

Similarly, it was urged at a Wenner-Gren symposium on the Middle Pleistocene
that its base be defined by the Matuyama-Brunhes reversal (Butzer and Isaac

1975, appendix 2). The same symposium proposed that the end of the Middle
Pleistocene, hence the base of the Upper Pleistocene, be defined at the
beginning of the Last Interglacial marine trasngression. Accordingly the base
of the Middle Pleistocene is so indicated on Table 10-1. But because the
precise dating of the commencement of the Last Interglacial marine transgression is difficult to place, and because numerous radiometric dates exist for
its culmination, ca. 125 ka BP, the lower boundary of the Upper Pleistocene
is placed at 128 ka BP (Table 10-1), which is the average date estimated for
Termination II of the ocean cores (Chapter 3). All three boundaries rest on
radiometric, or radiometric and palaeomagnetic criteria, which are independent
of the vagaries of biotal evolution and shift due to climatic change. All
three, moreover, are capable of global recognition, in the oceans and on the
continents.

CONTINENTAL - OCEANIC CORRELATION

Dating by means of carbon-14 has placed the events of the Last Glaciation on a
sound basis. It is true that some uncertainty continues to obtain beyond the
limit of that method, but the recent 'stretching' of the chronology, using
the Last Interglacial (125 ka BP) as control, appears to meet this contingency
adequately (Fig. 10-1). Correlation with the oxygen isotope record of the
oceans (stages 2 to 5d) seems firmly established. To what extent this should
be considered a model for earlier glaciations, however, is uncertain.
Proximity to the present preserved an unrivalled wealth of detailed evidence
unlikely to be forthcoming for earlier ones.

Middle Pleistocene correlations, to say nothing of Early Pleistocene ones
(Poag 1972. Zagwijn 1974) are fraught with difficulties and uncertainties.
Absence of a generally applicable radiometric dating technique makes the
correlation of continental and oceanic events tenuous. Part of the difficulty,
though it is a secondary consideration, arises because deep sea core boundaries (mid-points of isotopic extreme values), are not likely to be recorded
by definitive physical or biotal response on land. If glacials and interglacials are defined by 'mid-point' boundaries of the oxygen isotope stages,
it follows that the climatic changes to which they relate were already well
under way, being recorded on land by infinitely variable lithofacies and bio-
facies. Alternatively, if the boundaries are drawn at Terminations (Chapter
3), there is still no certainty that climate changes at such points. These
difficulties are persistent one because the mode for classification is a
climatic one - climatostratigraphic units. The difficulty is in relating
suitable criteria in rock units for inferred climatic ones on a basis meaningful for continental-oceanic correlation.

Logically the best boundaries would be those placed at maxima and minima in
the deep-sea cores : that is, reflecting periods of maximum glaciation, and
the temperature optimum of interglacials. Such events are more readily
recorded, and interpreted on land : for example, by climatic optima from
biotal data, and from palaeosols and culminations of marine transgressions
(Morrison 1965). Conventional procedures, however, are too deeply ingrained
to change, and the procedure outlined above does at least constitute a powerful means of correlation. Meanwhile the boundaries will continue to be
arbitrarily drawn, by definition almost, negating the climatic entities they
seek to define. On purely pragmatic grounds, tinged by expediency, the most
useful boundaries for the moment, are the deep sea core Terminations (not a
stratigraphic term), which Kukla (1975. 1977) correlates with the Loess
Marklines. The latter antedate thermal optima of interglacials by ca. 5,000

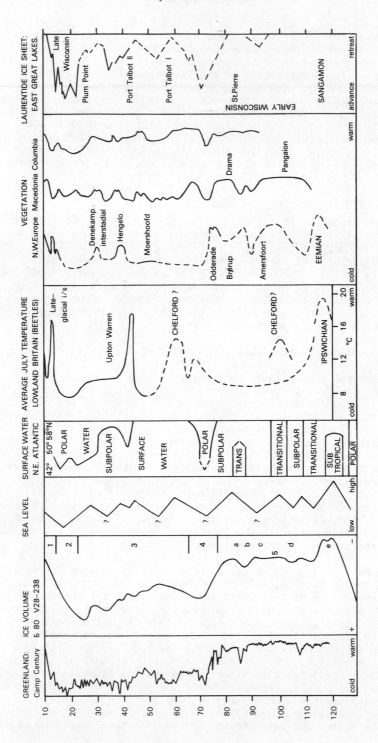

Fig. 10-1. Correlations of the Last Glaciation (from various sources). Sea levels from New Guinea

years, although this may be greater for earlier interglacials.

The terminology of Loess cycles, marine glacial cycles, and marine hemicycles
(Chapter 7) is in some respects potentially misleading, in much the same way
as Zeuner's terms became (Chapter 4), unless users of the data realise that
they refer to, and correlation is based on, specific units at a stratotype :
e.g. core K708-7 (Ruddiman and McIntyre 1976) for the glacial cycles. But
the absence of designated stratotypes for the Loess sequences is disturbing
for they are not based on continuous sections. This constitutes less of a
problem for 'workers' with such data, but for 'users' homotaxial misuse is
always possible.

Future redefinition may increase the number of formally recognised interglacials and glacials. A current dilemma is posed by stage 7 of the deep-sea
cores, subdivided into 7a, 7b and ·7c by Ninkovitch and Shackleton (1977), and
with continental ice sheets attaining 80% of their maximum size during stage
7b (Chapter 6). On this basis, ought not 7a and 7c be recognised as separate
interglacials? It may be that the high degree of resolution available for the
later stages, such as the Last Interglacial and Glaciation, will never be
possible for earlier ones. Hence different criteria might apply. But the
thought remains that Emiliani's (1955) original subdivision has conditioned
classification and may require revision.

In similar vein it is to be noted that the dilemma of the Last Interglacial is
not yet resolved to the satisfaction of all. Many regard stage 5e as the Last
Interglacial, followed by interstadials at ċa. 100 and 80 ka BP. As such it
conforms closely with present day global parameters (Dreimanis and Raukas
1975). But others regard 5a and 5c as sub-minima of the Last Interglacial
(Butzer 1976).

MEANS OF CORRELATION

Traditional means of dating and correlation on the continents are hopelessly
inadequate and are faced with new problems mostly beyond their resolving
capabilities. Almost exclusively, therefore, the burden for dating and
correlation falls on radiometric and geomagnetic methods. The evidence in
most regions, however, is not amenable to such methods. Other than in regions
of volcanic activity there are huge gaps in the Middle and Early Pleistocene
record. For the foreseeable future such regions will continue to pose extreme
problems. Procedure should, therefore, follow recommended stratigraphic
practice (Hedberg 1976), not least so that concepts and facts are clearly
understood by all. Units must be defined, or redefined, in lithostratigraphic,
biostratigraphic and chronostratigraphic terms, from local to provincial and
finally to standard scale levels. In such regions the use of a floating
chronology (Table 10-1) is desirable, wherein units are flexibly disposed
temporally, and are readily amenable to changing emphasis due to new discoveries or interpretations. Many outstanding problems in such regions can only
be resolved in the short term by detailed quadrangle and sheet mapping, aided
by section and subsurface logging.

This philosophy of *fixed points* or fixed events is not dissimilar to that
expounded by Kitts (1966) in that they mark the end of one period, or event,
and the beginning of the next. As more data becomes available the 'floating'
units can be tagged and placed within the larger ones.

On Table 10-1 uncertainty is denoted by italicised names. For northern Europe
the same stage names have been used at different stratigraphic horizons.

	Polarity	V28-238 stages	ka BP	K708-7 Cycles	Termina-tions	Marklines	Loess Cycles
HOLOCENE		1	—13—	A	—I—		
UPPER PLEISTOCENE	Blake	2					
		3					
		4					
		5a		B			B
		5b					
		5c					
	S	5d	—118—				
		5e	—128—		—II—	—II—	
MIDDLE PLEISTOCENE	E	6					
		7a		C			C
	H	7b					
	N	7c	—250—		—III—	—III—	
		8		D			D
	U	9	—350—		—IV—	—IV—	
		10		E			E
	R	11	—440—		—V—	—V—	
		12		F			F
	B	13	—500—		—VI—	—VI—	
		14		G			G
		15	—590—		—VII—	—VII—	
		16		H			H
		17	—640—		—VIII—	—VIII—	
		18					I
		19	—700—		—IX—	—IX—	
EARLY PLEISTOCENE		20					J
	M A T U Y A M A	21	—780—		—X—	—X—	
		22					K
		23	—900—		—XI—	—XI—	
	Jaramillo	24					
	Olduvai		—1.61 Ma—				

Table 10-1.   Correlation using a stratigraphic framework of polarity and isotopic (V28-238) subdivisions. Correlation with North America is premature, but a K/Ar and Fission Track dated sequence of tills, tephra and lake silts

Brno Terraces Fig. 9-1	Rhine Terraces Fig. 9-2	Karlich Fig. 9-3	Northern Europe	British Isles	Macedonia
CK 1	Low T	J	Weichsel	Devensian	
			Brørup	Chelford	
			Amersfoort	*Wretton*	
			Eemian	Ipswichian Trafalgar Sq	Pangaion
		Ja	Warthe		
	MT IV		Eemian	*Ipswichian Brundon ?*	Symvolon
CK 2		H	Saale Eemian	*Wolstonian Ipswichian Ilford*	
	MT III		Saale Holstein	*Anglian Hoxnian*	Lekanis Complex
		G	Elster Holstein	*Anglian Hoxnian*	
	MT II				Boz Dagh Complex
		F		*Cromerian*	
CK 3	MT I		Elster Cromerian	*Bestonian Pastonian*	Phalakron
		Bb C D E	Cromerian		
CK 4	HT	Ba	Cromerian		
CK 5			Menapian	*Baventian ?* *Antian* *Thurnian Ludhamian* *Baventian ?*	
			Waalian Eburonian		

covering all the Pleistocene occurs in Yellowstone National Park (Richmond 1976). The correlation of deep sea events with North America of Berggren and van Couvering 1974) based on the continental model, requires urgent revision.

REFERENCES

Aarseth, I and Mangerud, J. 1974 Younger Dryas end moraines between
    Hardangerfjorden and Sogefjorden, Western Norway, *Boreas*, 3, 3-22.
Aitken, M.J. 1974 *Physics and Archaeology*, Oxford.
Alt, D. and Brooks, H.K. 1965 Age of the Florida marine terraces, *J. Geol.*,
    73, 406-411.
American Commission on Stratigraphic Nomenclature 1961 Code of Stratigraphic
    nomenclature, *Bull. Am. Ass. Pet. Geol.*, 45, 646-665. (previous
    edition, 1933).
Anderson, S.T. 1966 Interglacial succession and lake development in Denmark,
    The Palaeobotanist, 15, 117-127.
Andrews, J.T. and Mahaffy, M.A.W. 1976 Growth rate of the Laurentide Ice
    Sheet and sea level lowering, *Quat. Res.*, 6, 167-184.
Antevs, E. 1955 Varve and radiocarbon chronologies appraised by pollen data,
    J. Geol., 63, 495-499.
Arrhenius, G. 1963 Pelagic sediments, in *The Sea* (3) (ed M.N. Hill), 655-727,
    London.
Averdieck, F.R. 1967 Die Vegetationsentwicklung des Eem-Interglazials und der
    Frün-würm-Interstadiale von Odderade/Schleswig-Holstein, *Fundamenta*
    B/2, 101-125.
Baker, C.A. 1976 Late Devensian periglacial phenomena in the upper Cam Valley,
    north Essex, *Proc. Geol. Ass.*, 87, 285-306.
Baker, V.R. 1974 Paleohydraulic interpretation of Quaternary alluvium near
    Golden, Colorado, *Quat. Res.*, 4, 94-112.
Bededict, J.B. 1976 Frost creep and gelifuction features: a review, *Quat.
    Res.*, 6, 55-76.
Berger, W.H. 1968 Planktonic foraminifera: selective solution and palaeo-
    climatic interpretation, *Deep Sea Research*, 15, 31-43.
Berger, W.H. 1970a Planktonic foraminifera: selective solution and the
    lysocline, *Marine Geol.*, 8, 111-138.
Berger, W.H. 1970b Biogenous deep-sea sediments: fractionation by deep sea
    circulation, *Bull. Geol. Soc. Am.*, 81, 1385-1402.
Berti, A.A. 1975 Palaeobotany of Wisconsinan Interstadials, Eastern Great
    Lakes Region, North America, *Quat. Res.*, 5, 591-620.
Birkeland, P. 1974 *Pedology, Weathering and Geomorphological Research*, Oxford.
Bishop, M.J. 1975 Earliest record of man's presence in Britain, *Nature,
    Lond.*, 253, 95-7.
Bishop, W.W. 1962 Pleistocene chronology in East Africa, *Adv. Sc.*, 18, 491-
    494.
Bishop, W.W. 1970 Discussion, *Proc. Geol. Soc. Lond.*, 1660, 380-382.
Bishop, W.W. and Miller, J.A. 1972 *Calibration of Hominoid Evolution*,
    Edinburgh.
Bishop, W.W. and Coope, G.R. 1977 The environmental history of Late glacial
    and early Postglacial times in south west Scotland. In *Studies in
    the Scottish Late glacial environment* (ed J.M. Gray and J.J. Lowe),
    London.
Black, R.F. 1976 Quaternary Geology of Wisconsin and Contiguous Upper
    Michigan. In *Quaternary Stratigraphy of North America* (ed W.C.
    Mahaney), 93-117, Stroudsburg, Penn.
Blasing, T.J. and Fritts, H.C. 1976 Reconstructing past climatic anomalies in
    the North Pacific and Western North America from Tree ring data,
    Quat. Res., 6, 563-580.
Bleur, N.K. 1974 Buried till ridges in the Fort Wayne area, Indiana, and
    their regional significance, *Bull. Geol. Soc. Am.*, 85, 917-920.

Bloom, A.L. 1967 Pleistocene shorelines: a new test of isotasy, *Bull. Geol. Soc. Am.*, 78, 1477-94.

Bloom, A.L. 1970 Paludal stratigraphy of Truk, Ponape and Kusaie, Eastern Caroline Islands, *Bull. Geol. Soc. Am.*, 81, 1895-1904.

Bloom, A.L. 1971 Glacial-eustatic and isostatic controls of sea level since the Last Glaciation. In *The Late Cenozoic Glacial Ages* (ed K.K. Turekian), 355-379, New Haven.

Bloom, A.L., Broecker, W.S., Chappell, J.S., Matthews, R.K. and Mesolella, K.J. 1974 Quaternary sea-level fluctuations on a tectonic coast: New Th^{230}/U^{234} dates from the Huon Peninsula, New Guinea, *Quat. Res.*, 4, 185-205.

Boellstorff, J. 1973 Correlating and dating some older "Pleistocene" tills in the midcontinent, *Geol. Soc. Am. Abs.*, 5(4), 301.

Borchardt, G.A., Harward, M.E. and Schmitt, R.A. 1971 Correlation of Volcanic Ash Deposits by Activation Analysis of Glass Separates, *Quat. Res.*, 1, 247-260.

Boreas 1977 Symposium on till genesis, *Boreas,* 6(2), 71-228.

Boulton, G.S. 1972 Modern arctic glaciers as depositional models for former ice sheets, *J. geol. Soc. Lond.*, 182, 361-393.

Boulton, G.S. 1975 Processes and patterns of subglacial sedimentation: a theoretical approach. In *Ice Ages: Ancient and Modern* (eds A.E. Wright and F. Moseley), 7-42, Geol. J. Spec. Issue No. 6, Liverpool.

Boulton, G.S. 1977 A multiple till sequence formed by a late Devensian Welsh ice-cap, Glanllynnau, Gwynedd, *Cambria,* 4, 10-31.

Boulton, G.S. and Worsley, P.W. 1965 Late Weichselian glaciation of the Cheshire-Shropshire Basin, *Nature, London*, 207, 704-706.

Boulton, G.S., Jones, A.S., Clayton, K.M. and Kenning, M.J. 1977 A British Ice Sheet Model and patterns of glacial erosion and deposition in Britain. In *British Quaternary Studies* (ed F.W. Shotton), 231-246, Oxford.

Bowen, D.Q. 1969 A new interpretation of the Pleistocene succession in the Bristol Channel, *Proc. Ussher Soc.*, 2, 89.

Bowen, D.Q. 1973a The Pleistocene history of Wales and the borderland, *Geol. J.*, 8, 207-224.

Bowen, D.Q. 1973b The Pleistocene succession of the Irish Sea, *Proc. Geol. Ass.*, 84, 249-272.

Bowen, D.Q. 1974 The Quaternary of Wales. In *The Upper Palaeozoic and post Palaeozoic rocks of Wales* (ed T.R. Owen), 373-426, Cardiff.

Bowen, D.Q. 1977 The coast of Wales. In *The Quaternary History of the Irish Sea,* Spec. No. Geol., No. 7, 223-256.

Bretz, J.H., Smith, H.T.U. and Neff, G.E. 1956 Channeled Scabland of Washington; New Data and Interpretations, *Bull. Geol. Soc. Am.,* 67, 957-1050.

Brewer, R., Crook, K.A.W. and Speight, J.G. 1970 Proposal for soil strati-graphic units in the Australian stratigraphic code, *J. Geol. Soc. Aust.*, 17, 103-111.

Bristow, C.R. and Cox, F.C. 1973 The Gipping Till; a reappraisal of East Anglian glacial stratigraphy, *J. geol. Soc. Lond.*, 129, 1-37.

Broecker, W.S. 1965 Isotope geochemistry and the Pleistocene climate record. In *The Quaternary of the United States* (eds H.E. Wright and D.G. Frey), 737-53, Princeton.

Broecker, W.S. and Kaufman, A. 1965 Radiocarbon chronology of Lake Lahontan and Lake Bonneville, *Bull. Geol. Soc. Am.*, 76, 537-66.

Broecker, W.S., Thurber, D.L., Goddard, J., Ku, T.L., Matthew, R.K. and Mseollela, K.J. 'Milankovitch hypothesis supported by precise dating of coral reefs and deep-sea sediments, *Science,* 159, 297-300.

Broecker, W.S. and van Donk, J. 1970 Insolation changes, ice volumes, and the
        O^{18} record in deep sea cores, *Rev. Geophysics and Space Phys.*, 168-
        198.

Broecker, W.S. and Bender, M.L. 1972 Age determinations on marine strandlines.
        In *Calibration of Hominoid Evolution* (eds W.W. Bishop and J.A.
        Miller), 19-38, Edinburgh.

Brunnacker, K. 1975 The Mid-Pleistocene of the Rhine Basin. In *After the
        Australopithecines* (eds K.W. Butzer and G.Ll. Isaac), 189-224,
        The Hague.

Brunnacker, K. 1977 in Excursion Guidebook C17: Southern Shores of the North
        Sea (ed D.Q. Bowen), *X INQUA Congress.*

Brunnacker, M and Brunnacker, B 1962 Weitere Funde pleistozäner Mollusken-
        faunen bei München, *Eisz. und Gegen*, 13, 129-137.

Büdel, J. 1949 Die räumliche und zeitliche Gliederung Eiszeitklima, *Die
        Naturwissenschafen*, 36, 105-112, 133-139.

Burleigh, R. 192 Carbon-14 Dating, *Studies in Speleology*, 2, 176-190.

Butzer, K.W. 1971 *Environment and archeology – an ecological approach to
        prehistory*, London.

Butzer, K.W. 1975 Geological and ecological perspectives on the Middle
        Pleistocene. In Butzer, K.W. and Isaac, G.L., *After the Australo-
        pithecines*, 857-874, The Hague.

Butzer, K.W. 1976 Pleistocene Climates. In Ecology of the Pleistocene,
        Geoscience and Man, 13, 27-44.

Butzer, K.W. 1977 Environment, Culture and Human Evolution, *American
        Scientist*, 65, 572-584.

Butzer, K.W. and Isaac, G.Ll. (eds) 1975 *After the Australopithecines*, The
        Hague.

Butzer, K.W. and Cuerda, J. 1962 Coastal stratigraphy of southern Mallorca
        and its implication for the Pleistocene chronology of the
        Mediterranean Sea, *J. Geol.*, 70, 398-416.

Catt, J.A. 1977 Loess and Coversands. In *British Quaternary Studies*, (ed
        F.W. Shotton), 221-230, Oxford.

Chamberlain, T.C. 1895 Glacial phenomena of North America. In *The Great Ice
        Age*, James Geikie, 724-755, New York.

Chappell, J.M.A. 1974a Late Quaternary Glacio- and Hydro-isostasy on a
        layered earth, *Quat. Res.*, 4, 405-428.

Chappell, J.M.A. 1974b Geology of coral terraces, Huon Peninsula, New
        Guinea: a study of Quaternary tectonic movements and sea level
        changes, *Bull. Geol. Soc. Am.*, 85, 553-570.

Chappell, J.M.A. and Polach, H.A. 1972 Some effects of partial recrystallis-
        ation on 14C dating Late Pleistocene Corals and Molluscs, *Quat. Res.*,
        2, 244-252.

Charlesworth, J.K. 1957 *The Quaternary Era* (2 vols), London.

Chorley, R.J., Dunn, A.J. and Beckinsale, R.P. 1964 *The history of the study
        of landforms, or the development of Geomorphology*, London.

Clapperton, C.M. and Sugden, D. 1972 The Aberdeen and Dinnet glacial limits
        reconsidered. In *North-east Scotland Geographical Essays* (ed C.M.
        Clapperton), 5-11, Aberdeen.

Clark, J.D. 1957 *3rd Pan African Congress on Prehistory*, London.

Clark, J.D. 1965 The later Pleistocene cultures of Africa, *Science*, 150,
        833-847.

Clayton, K.M. 1977 River Terraces. In *British Quaternary Studies* (ed F.W.
        Shotton), 153-168, Oxford.

CLIMAP 1976 The surface of ice age earth, *Science*, 191 (4232).

Cooke, H.B.S. 1972 Evolution of mammals on southern continents, 2, the fossil
        mammal fauna of Africa, *Q. Rev. Biol.*, 43, 234-264.
Coope, G.R. 1959 A Late Pleistocene insect fauna from Chelford, Cheshire,
        Proc. R. Soc. Lond., B 151, 70-86.
Coope, G.R. 1975 Climatic fluctuations in north-west Europe since the Last
        Interglacial, indicated by fossil assemblages of Coleoptera. In
        Ice Ages ancient and modern (ed A.E. Wright and F. Moseley), 153-168,
        Geol. J. Spec. Issue, 6, Liverpool.
Coope, G.R. 1977 Fossil coleopteran assemblages as sensitive indicators of
        climatic changes during the Devensian (Last)cold stage, *Phil. Trans.
        R. Soc. Lond.*, B 280, 313-340.
Coope, G.R. and Pennington, W. 1977 The Windermere Interstadial of the Late
        Devensian, *Phil. Trans. R. Soc.*, B 208.
Cox, A., Doell, R.R. and Dalrymple, G.B. 1963 Geomagnetic polarity epochs and
        Pleistocene geochronometry, *Nature London*, 198, 1049-1051.
Craig, H. 1961 Standard for reporting concentrations of deuterium and oxygen-
        18 in natural waters, *Science*, 133, 1833.
Creer, K.M., Gross, D.L. and Lineback, J.A. 1976 Origin of regional geo-
        magnetic variations recorded by Wiconsinan and Holocene sediments
        from Lake Michigan, U.S.A., and Lake Windermere, England, *Bull.
        Geol. Soc. Am.*, 87, 531-540.
Cushing, E.J. 1967 Late-Wisconsin Pollen Stratigraphy and the Glacial
        Sequence in Minnesota. *In Quaternary Palaeoecology* (eds Cushing, E.J.
        and Wright, H.E.), 59-88, New Haven.
Dalrymple, G.B. 1972 Potassium-argon dating of geomagnetic reversals and
        North American glaciations. In *Calibration of Hominoid Evolution*
        (eds W.W. Bishop and J.A. Miller), 107-134, Edinburgh.
Daly, R.A. 1925 Pleistocene changes of level, *Am. J. Sc.*, 10, 281-313.
Daly, R.A. 1934 *The Changing World of the Ice Age*, Yale.
Damon, P.E., Long, A. and Wallick, E.I. 1973 Dendrochronologic calibration
        of the carbon-14 time scale, *Proc. 8th Int. Conf. Radiocarbon
        Dating* (New Zealand), 18-25.
Damuth, J.E. and Fairbridge, R.W. 1970 Arkosic sands of the Last Glacial
        stage in the tropical Atlantic off Brazil, *Bull Geol. Soc. Am.*, 81,
        189-206.
Dansgaard, W.S. and Tauber, H., 1969 Glacial oxygen-18 content and Pleistocene
        ocean temperatures, *Science*, 166, 499-502.
Dansgaard, W.S., Johnsen, S.J., Møller, J. and Langway, C.C. 1969 One thou-
        sand centuries of climatic record from Camp Century on the Greenland
        ice sheet, *Science*, 166, 377.
Dansgaard, W.S., Johnsen, S.J., Clausen, H.B. and Langway, C.C. 1971 Climatic
        record revealed by the Camp Century Ice Core. In *The Late Cenozoic
        glacial ages* (ed K.K. Turekian), 37-56, New Haven.
Davis, M.B. 1969 Climatic changes in Connecticut recorded by pollen deposit-
        ion at Rogers Lake, *Ecology*, 50, 409.
Davis, M.B. 1976 Pleistocene biogeography of temperate deciduous forests. In
        Ecology of the Pleistocene, *Geoscience and Man*, 13, 13-26.
Davis, M.B. and Deevey, E.S. 1964 Pollen accumulation rates: estimates from
        Late-glacial sediment of Rogers Lake, *Science*, 145, 1293-1295.
Davis, R.B. and Webb, T. 1975 The contemporary distribution of pollen in
        eastern North America: a comparison with vegetation, *Quat. Res.*,
        5, 395-434.
Deevey, E.S. 1965 Pleistocene Nonmarine Environments. In *The Quaternary of
        the United States* (eds H.E. Wright and D.G. Grey), 643-654,
        Princeton.

Donk, J. van 1976 An ^{18}O record of the Atlantic Ocean for the entire Pleistocene. In *Investigation of Late Quaternary Paleoceanography and Paleoclimatology* (eds R.M. Cline and J.D. Hays), 147-164, *Mem. geol. Soc. Am.*, 145.

Donner, J.J. 1970 Land/sea-level changes in Scotland. In *Studies in the vegetational history of the British Isles* (eds D. Walker and R.G. West), 23-39, Cambridge.

Donner, J.J., Junger, H. and Vasari, Y. 1971 The hard-water effect on radio-carbon measurements of samples from Saynajalampi, north-east Finland, *Comm. Physico-Math.*, 41, 307-310.

Donner, J.J. and Junger, H. 1973 The effect of re-deposited organic material on radiocarbon measurements of clay samples from Somero, south-western Finland, *Geol. Foren. Stockh. Forh.*, 95, 267-268.

Donovan, D.T. 1962 Sea levels of the Last Glaciation, *Bull. Geol. Soc. Am.*, 73, 1297-1298.

Dort, W., Jr. 1972 Stadial subdivisions of early Pleistocene glaciations in central United States - a developing chronology, *Boreas*, 1, 55-62.

Dreimanis, A. 1976 (Terasmae, J. and Dreimanis, A), Quaternary Stratigraphy of Southern Ontario. In *Quaternary Stratigraphy of North America*, (ed W.C. Mahaney), 51-64, Stroundsburg, Penn.

Dreimanis, A., and Goldthwait, R.P. 1973 Wiconsin glaciation in the Huron, Erie and Onatio Lobes, *Mem. geol. Soc. Am.*, 136, 71-106.

Dreimanis, A. and Raukas, A. 1973 Did Middle Wisconsin, Middle Weichselian and Their Equivalents Represent an Interglacial or an Interstadial Complex in the Northern Hemisphere? In *Quaternary Studies* (eds R.P. Suggate and M.M. Cresswell), Wellington, N.Z.

Droste, J.B., Rubin, M. and White, G.W. 1959 Age of marginal drift at Corry, northwestern Pennsylvania, *Science*, 130, 1760.

Duphorn, K., Grube, F., Meyer, K.D., Streif, H. and Vinken, R. 1973 Pleisto-cene and Holocene, *Eisz. u. Gegenw.*, 23/24, 222-250.

Duplessy, J.C., Lalou, C. and Vinot, A.C. 1970 Differential isotopic fract-ionation in benthic foraminifera and palaeotemperatures reassessed, *Science*, 168, 250-251.

Easterbrook, D.J. and Othberg, K. 1976 Palaeomagnetism of Pleistocene sedi-ments in the Puget Lowland, Washington. In *Quaternary Glaciations in the Northern Hemisphere* (eds D.J. Easterbrook and V. Sibrava), 189-207, Prague 1976.

Eberl, B. 1930 *Die Eiszeitenfolge im nördlichen Alpenvorlande*, Augsburg.

Emiliani, C. 1954 Depth habitats of some species of pelagic foraminifera as indicated by oxygen isotope ratios, *Am. J. Sci.*, 252, 149.

Emiliani, C. 1955 Pleistocene temperatures, *J. Geol.*, 63, 538-78.

Emiliani, C. 1966 Palaeotemperature analysis of Carribbean cores and a generalized temperature curve for the last 425000 years, *J. Geol.*, 74, 109-26.

Emiliani, C. 1971 The amplitude of Pleistocene climatic cycles at low latit-udes and the isotopic composition of glacial ice. In *The Late Cenozoic Glacial Ages* (ed K.K. Turekian), 183-197, Yale.

Emiliani, C. and Flint, R.F. 1963 The Pleistocene record. In *The Sea* (vol. 3) (ed M.N. Hill), Wiley, New York.

Emiliani, C. and Shackleton, N.J. 1974 The Brunhes Epoch: Palaeotemperatures and geochronology, *Science*, 183, 511-514.

Epstein, S., Buchsbaum, R., Lowenstam, H.A. and Urey, H.C. 1953 Revised carbonate-water isotopic temperature scale, *Bull. geol. Soc. Am.*, 64, 1315-26.

Epstein, S., Sharp, R.P., and Gow, A.J. 1970 Antarctic ice sheet isotope analysis of Byrd Station cores and interhemispheric climatic implic-ations, *Science*, 168, 1570-2.

Ericson, D.B. and Wollin, G. 1968 Pleistocene climates and chronology in deep-sea sediments, *Science*, 162, 1227-34.

Evans, P. 1971 The Phanerozoic time-scale. A supplement. Part 2. Towards a time-scale, *Geol. Soc. Lond. Spec. Pub.*, 5.

Evenson, E.B., Farrand, W.R., Eschman, D.F., Mickleson, D.M. and Maher, L.J. 1976 Greatlakean Substage: A Replacement for Valderan Substage in the Lake Michigan Basin, *Quat. Res.*, 6, 411-424.

Faegri, K. and Iversen, J. 1975 *Textbook of pollen analysis* (1st ed 1964), Copenhagen.

Fairbridge, R.W. 1961 Eustatic changes in sea-level, *Physics and Chemistry of The Earth*, 4, 99-185.

Fairbridge, R.W. 1968 Quaternary Period. In *The Encyclopedia of Geomorphology* (ed R.W. Fairbridge), 912-928, New York.

Fairbridge, R.W. 1971 Quaternary shoreline problems at INQUA, *Quaternaria*, 15, 1-17.

Fairbridge, R.W. 1972 Climatology of a glacial cycle, *Quat. Res.*, 2, 283-302.

Fairbridge, R.W. 1973 Friends of the Mediterranean Quaternary visit type sections, *Geotimes*, 24-26 (Nov).

Fairbridge, R.W. 1976 Effects of Holocene climatic change on some tropical geomorphic processes, *Quat. Res.*, 6, 529-556.

Farrand, W.R., Zahner, R. and Benninghof, W.S. 1969 Cary-Port Huron Interstade: evidence from a buried Bryophyte bed, Cheboygan County, Michigan, *Geol. Soc. Am. Spec. Paper*, 123, 249-262.

Ferguson, C.W. 1972 Dendrochronology of bristlecone pine prior to 4000 B.C. *Proc. 8th Int. Conf. on Radiocarbon dating* (eds. T.A. Rafter and T. Grant-Taylor), 17-26, Wellington, N.Z.

Fitch, J.F. 1972 Selection of suitable material for dating and the assessment of geological error in potassium-argon age determination. In *Calibration of Hominoid Evolution* (ed W.W. Bishop and J.A. Miller), 77-91, Edinburgh.

Fitch, J.F. and Miller, J.A. 1970 Radioisotopic age determination of Lake Rudolf artefact site, *Nature London*, 226, 226-228.

Fitch, J.F., Hooker, P.J. and Miller, J.A. 1976 Argon-40/argon-39 dating of the KBS tuff in Koobi Fora Formation East Rudolf Kenya, *Nature London*, 263, 740-744.

Fleischer, R.L. and Hart, H.R. 1972 Fission track dating techniques and problems. In *Calibration of Hominoid Evolution* (ed W.W. Bishop and J.A. Miller), 138-70.

Flemming, N.C. and Roberts, D.G. 1973 Tectono-eustatic changes in sea level and seafloor spreading, *Nature London*, 243, 19-22.

Flint, R.F. 1957 *Glacial and Pleistocene Geology*, New York.

Flint, R.F. 1959 On the basis of Pleistocene correlation in East Africa, *Geol. Mag.* 96, 265-284.

Flint, R.F. 1971 *Glacial and Quaternary Geology*, New York.

Flint, R.F. 1974 Three theories in Time, *Quat. Res.*, 4, 1-8.

Flohn, H. 1974 Background of a geophysical model of the initiation of the next glaciation, *Quat. Res.*, 4, 385-404.

Forbes, E. 1846 On the connection between the distribution of the existing fauna and flora of the British Isles ...., *Geol. Surv. Gt. Brit. Mem.*, 1, 336-342.

Francis, E.A. 1975 Glacial sediments: a selective review. In *Ice Ages: Ancient and Modern* (eds A.E. Wright and F. Moseley), *Geol. J. Spec. Issue No. 6*, Liverpool.

French, H.M. 1976 *The Periglacial Environment*, London.

Frenzel, B. 1973 *Climatic fluctuations of the Ice Age*, Cleveland.

Frey, D.G. 1964 Remains of animals in Quaternary lake and bog sediments and their interpretation, *Arch. Hydrol., suppl. Ergebn. der Limnol,* 2, 1-114.

Fritts, H.C. 1965 Dendrochronology. In *The Quaternary of the United States* (eds H.E. Wright and Frey, D.G.), 871-879, Princeton,

Frye, J.C. 1973 Pleistocene succession of the central interior United States, *Quat. Res.,* 3, 275-283.

Geer, de G. 1940 Geochronologia Sueccia principles, *K. Svenska Vetensk-Akad. Handl.* Ser. 3, 18(6).

George, T.N. 1970 Discussion, *Proc. Geol. Soc. Lond.,* 1660, 378-380.

George, T.N. 1976 Charles Lyell: The Present is the Key to the Past, *Philosoph. Jl.,* 13, 3-24.

Gignoux, M. 1950 *Stratigraphic Geology,* Paris.

Godwin, H. 1956 *The history of the British flora,* Cambridge.

Godwin, H. 1962 Half-life of radiocarbon, *Nature London,* 195, 984.

Godwin, H. and Willis, E.H. 1958 Radiocarbon dating of the Eustatic Rise in Ocean Level, *Nature London,* 181, 1518.

Goldthwait, R.P. (ed) 1971 *Till: a symposium,* Columbus, Ohio.

Gordon, A.L. 1975 General ocean circulation. In *Numerical models of ocean circulation* (Nat. Acad. Sci.), 39-53, Washington.

Graul, H., 1973 Foreland of the Alps: lithostratigraphy, palaeopedology and Geomorphology, *Eisz. u. Gegnw.,* 23/24, 268-280.

Gray, J.M. 1974 The main rock platform of the Firth of Lorne, Western Scotland, *Trans. Inst. Br. Geogr.,* 61, 81-99.

Gruhn, R., Bryan, A.L. and Moss, A.J. 1974 A contribution of the Pleistocene chronology in southeast Essex, England, *Quat. Res.,* 4, 53-75.

Hammen, T. Van Der, Wijmstra, T.A. and Zagwijn, W.H. 1971 The floral record of the Late Cenozoic of Europe. In *The Late Cenozoic glacial ages* (ed K.K. Turekin), 391-424.

Handa, S. Moore, P.D. 1976 Studies in the vegetational history of mid-Wales IV, *New Phytol.,* 77, 203-233.

Hare, P.E. and Taylor, R.E. 1970 Amino acid ages: comparison with radiometric determinations and potential applications to archaeological materials, *AMQUA abstracts,* Bozeman meeting.

Hay, R.L. 1973 Lithofacies and environments of Bed I, Olduvai Gorge, Tanzania, *Quat. Res.,* 3, 541-560.

Hays, J.D., Saito, T., Opdyke, N.D. and Burckle, L.H. 1969 Pliocene-Pleistocene sediments of the Equatorial Pacific: their palaeomagnetic, biostratigraphic and climatic record, *Bull. Geol. Soc. Am.,* 80, 1481-1514.

Hays, J.D. and Moore, T.C. 1973 CLIMAP Program, *Quat. Res.,* 3, 1-2.

Hays, J.D., Imbrie, J., and Shackleton, N.J. 1976 Variations in the earth's orbit: pacemaker of the ice ages, *Science,* N.Y., 194, 1121-32.

Haq, B.U., Berggren, W.A. and Couvering, J.A. van 1977 Corrected age of the Pliocene boundary, *Nature London,* 269, 483-488.

Hecht, A.D. 1973 Faunal and Oxygen Isoptopic Palaeotemperatures and the amplitude of Glacial/Interglacial Temperature changes in the Equatorial Atlantic, Caribbean Sea and Gulf of Mexico, *Quat. Res.,* 3, 671-690.

Hecht, A. 1976 The oxygen isotope record of Foraminifera in deep sea sediments, *Foraminifera,* 2, 1-43.

Hedberg, H.D. (ed) 1976 *International Stratigraphic Guide,* New York.

Heinselman, M.L. 1973 Fire in the Virgin Forests of the Boundary Waters Cance Area, Minnesota, *Quat. Res.,* 3, 329-382.

Heusser, L. and Balsam, W.L. 1977 Pollen distribution in the Northeast Pacific Ocean, *Quat. Res.,* 7, 45-62.

Hey, R.W. 1971 Quaternary shorelines of the Mediterranean and Black Seas, *Quaternaria*, 15, 273-284.

Hibbard, C.W., Ray, D.E., Savage, D.E., Taylor, D.W. and Guilday, J.E. 1965. Quaternary Mammals of North America. In *The Quaternary of the U.S.A.* (ed H.E. Wright and D.G. Frey), 509-525, Princeton.

Hollin, J.T. 1965 Wilson's theory of ice ages, *Nature London*, 208, 12-16.

Hollin, J.T. 1976 Thames interglacial sites, Ipswichian sea levels and Antarctic ice surges, *Boreas*, 6, 33-52.

Hooten, J.E. 1972, cited by Johnson, W.H. 1976 Quaternary stratigraphy in Illinois. In *Quaternary Stratigraphy of North America* (ed W.C. Mahaney), Stroudsburg, Penn.

Horrie, S. 1976 *Paleolimnology of Lake Biwa and the Japanese Pleistocene*, vol. 4, 836 pp.

Hurford, A.J., Gleadow, A.J.W. and Naeser, C.W. 1976 Fission track dating from the KBS tuff, East Rudolf, Kenya, *Nature London*, 263, 738-740.

Imbrie, J.C. and Kipp, N.G. 1971 A new micropalaeontological method for Quantitative palaeoclimatology: application to a Late Pleistocene Caribbean Core. In *The Late Cenozoic Glacial Ages* (ed K.K. Turekian), 71-182, New Haven.

Imbrie, J., van Donk, J. and Kipp, N.G. 1973 Paleoclimatic investigation of a Late Pleistocene Caribbean Deep-Sea Core: Comparison of Isotopic and Faunal Methods, *Quat. Res.*, 3, 10-38.

Isaac, G.Ll. 1975 Sorting out the Muddle in the Middle: An Anthropologist's Post-Conference Appraisal. In *After the Australopithecines* (eds K.W. Butzer and G.Ll. Isaac), 875-888, The Hague.

Iversen, J. 1947 Plantevaekst, Dyreliv og klima i det senglaciale Danmark, *Geol. Foren. Stockholm Forh.*, 69,

Iversen, J. 1958 The bearing of glacial and interglacial epochs on the formation and extinction of plant taxa, *Uppsala Univ. Arsskrift*, 210.

Ives, J.D., Andrews, J.T. and Barry, R.G. 1975 Growth and decay of the Laurentide Ice Sheet and comparisons with Fenno-Scandinavia, *Die Naturwissenschaften*, 62, 118-125.

Izett, G.A., Wilcox, R.E., Powers, H.A. and Desborough, G.A. 1970 The Bishop Ash Bed, a Pleistocene marker bed in the Western United States, *Quat. Res.*, 1, 121-132.

Izett, G.A., Wilcox, R.E. and Borchardt, G.A. 1972 Correlation of a volcanic ash bed in Pleistocene Deposits near Mount Blanco, Texas, with the Guaje Pumice Bed of the Jemez Mountains, New Mexico, *Quat. Res.*, 2, 554-578.

Jardine, W.G., Dickson, J.H. and price, R.J. 1976 Three late-Devensian sites in West-central Scotland, *Nature London*, 262, 43-44.

Järnefors, B. 1963 Lervarvskronologien och isrecession i östra Mellansverige, *Sver. geo. unders*. C 594, 67 pp.

Jelgersma, S. 1966 Sea level changes in the last 10,000 years. In *Royal Meteorological Soc. Intern. Symposium on World Climate From 8000-0 B.C.*, 54-69.

Jessen, K. and Milthers, V. 1928 Stratigraphical and palaeontological studies of interglacial freshwwater deposits in Jutland and north-west Germany, *Danm. Geol. Unders.*, II Raekke, No. 48.

Johnsen, S.J., Dansgaard, W., Clausen, H.B. and Langway, C.C. 1972 Oxygen isotope profiles through the Antarctic and Greenland Ice Sheets, *Nature London*, 235, 429-434.

Johnson, W.H. 1976 Quaternary Stratigraphy in Illinois: Status and Current Problems. In *Quaternary Stratigraphy of North America* (ed W.C. Mahaney), 161-198, Stroudsburg, Penn.

de Jong, J.D. 1967 The Quaternary of the Netherlands. In *The Quaternary*, 2 (ed K. Rankama), New York.

Karrow, P.F. and Anderson, T.W. 1975 Palynological Study of Lake Sediment Profiles from Southwestern New Brunswick: Discussion, *Canadian J. Earth Sc.*, 12, 1808-1812.

Kaufman, A., Broecker, W.S., Ku, T.L., Thurber, D.L. 1971 The status of U-series methods of mollusk dating, *Geochimica et Cosmochimica Acta*, 35, 1155-1183.

Keihlack, K. 1926 Das Quartär. In *Grundzüge der Geologie*, Vol. 2 (ed Salmon), Stuttgart.

Kennett, J.P. and Huddleston, P. 1972 Late Pleistocene palaeoclimatology, foraminiferal biostratigraphy and tephrochronology, western Gulf of Mexico, *Quat. Res.*, 2, 38-69.

Kennett, J.P. and Shackleton, N.J. 1975 Laurentide ice sheet meltwater recorded in Gulf of Mexico deep sea cores, *Science*, N.Y., 188, 147-50.

Kerney, M.P. 1977 British Quaternary Nonmarine Mollusca: a brief review. In *British Quaternary Studies* (ed F.W. Shotton), 31-42, Oxford.

Kerney, M.P., Brown, E.H. and Chandler, T.J. 1964 The Late-glacial and Post-glacial history of the chalk escarpment near Brook, Kent, *Phil. Trans. R. Soc.*, B 248, 135-204.

Kershaw, A.P. 1974 A long continuous pollen sequence fron north-eastern Australia, *Nature*, 251, 222.

Kidson, C. 1977 The coast of south west England. In *The Quaternary History of the Irish Sea* (eds C. Kidson and M.J. Tooley), 257-298.

Kidson, C. and Heyworth, A. 1976 The Quaternary deposits of the Somerset Levels, *Q. Jl. Eng. Geol.*, 9, 217-235.

Kitts, D.B. 1966 Geologic Time, *J. Geol.*, 74, 127-146.

Kubiena, W.L. 1956 Zur Mikromorphologie, Systematik und Entwicklung der rezenten und fossilen Lössboden, *Eisz. u. Gegenw.*, 7, 102-112.

Kukla, G.J. 1970 Correlation between loesses and deep-sea sediments, *Geol. Foren Stockholm Förh.*, 92, 148-180.

Kukla, G.J. 1975 Loess stratigraphy of Central Europe. In *After the Australopithecines* (ed K.W. Butzer and G.Ll. Issac), 99-188, The Hague.

Kukla, G.J. 1977 Pleistocene Land-Sea Correlations. I. Europe, *Earth Sc. Rev.*, 13, 307-374.

Kukla, G.J., Mathews, R.K. and Mitchell, J.M. 1972 The end of the present interglacial, *Quat. Res.*, 2, 261-269.

Kukla, G.J. and Nakagawa, H. 1977 Late Cenozoic Magnetostratigraphy. Comparisons with Bio-Climato and Lithozones, *Quat. Res.*, 7, 283-293.

Kurten, B. 1968 *Pleistocene mammals of Europe*, London.

Langford-Smith, T. and Thom, B.G. 1969 New South Wales Coastal Morphology, *J. geol. Soc. Aust.*, 16, 572-580.

Leakey, L.S.B. 1952 (ed) *1st Pen-African Congress on Prehistory*, Oxford.

Leonard, A.B. *et al* 1971 Illinoian and Kansan molluscan faunas of Illinois, *Ill. Geol. Surv. Circ.*, 461, 23 pp.

Leverett, F. 1898 The weathered zone (Sangamon) between the Iowan loess and Illinoian till sheet, *J. Geol.*, 6, 171-181.

Leverett, F. 1899 The Illinois glacial lobe, *U.S. Geol. Surv. Mon.*, 38, 817 pp.

Libby, W.F. 1955 *Radiocarbon dating*, 2nd ed, Chicago.

Lidz, L. 1966 Deep sea biostratigraphy, *Science*, 154, 1448.

Lineback, J.A. 1975 Glacial landforms on Wisconsinan and Illinoian drift in east-central Illinois mapped from Skylab photographs, *Geol. Soc. Am. Abs.*, 7, 809.

Lozek, V. 1964 *Quatarmollusken der Tschecoslowakei*, Rozpravy Ustredniho ustau geologickeho, 31, Prague.

Lozek, V. 1969 Paläontohische Charakteristik der Löss-Serien. In *Periglacial-zone, Löess und Palaolithikum der Tscechoslowakei* (eds J. Demek and J. Kukla), 43-60, Brno.

Lozek, V. 1972 Holocene interglacial in Central Europe and its Land snails, *Quat. Res.*, 2, 327-334.

Lundelius, E.L. 1976 Veterbrate Palaeontology of the Pleistocene: An Overview, in the Ecology of the Pleistocene, *Geoscience and Man*, 13, 45-59.

Lundquist, J. 1975 Ice recession in central Sweden, and the Swedish Time Scale, *Boreas*, 4, 47-54.

Luttig, G. 1959 Eiszeit-Stadium-Phase-Staffel Eine nomenklatorische Betrachtung, *Geol. Jahrb.*, 76, 235-268.

Luttig, G., Paepe, R., West R.G. and Zagwijn, W.H. 1969 *Key to the interpretation and nomenclature of Quaternary stratigraphy*, Hannover, 46 pp.

Luz, B. and Shackleton, N.J. 1975 CaCO3 solution in the tropical East Pacific during the past 130,000 years, *Cushman Foundn. Foram. Res. Spec. Pubn.*, 13, 142-150.

Lyell, C. 1839 *Nouveaux elements de geologie*, Paris.

McGraw, J.D. 1975 Quaternary Airfall Deposits of New Zealand. In *Quaternary Studies* (eds R.P. Suggate and M.M. Cresswell), Wellington, N.Z.

Machida, H. 1975 Pleistocene Sea-Level changes of South Kanto, Central Japan, Analysed by Tephrochronology. In *Quaternary Studies* (eds R.P. Suggate and M.M. Cresswell), Wellington, N.Z.

McIntyre, A. and Ruddiman, W.F. 1972 Northeast Atlantic Post Eemian Paleo-oceanography: a Predictive Analog of the Future, *Quat. Res.*, 2, 350-354.

McIntyre, A., Kipp, N.G. *et al* 1976 Glacial North Atlantic 18,000 years ago: a CLIMAP reconstruction, *Geol. Soc. Am. Mem.*, 143, 43-76.

Madgett, P.A. 1975 Reinterpretation of Devensian till stratigraphy of eastern England, *Nature London*, 253, 105-107.

Maenaka, K., Yokoyama, T., and Ishida, S. 1977 Paleomagnetic stratigraphy and biostratigraphy of the Plio-Pleistocene in the Kinki District Japan, *Quat. Res.*, 7, 341-362.

Mangerud, J., Andersen, S.T., Berglund, B.E. and Donner, J.J. 1974 Quaternary stratigraphy of Norden, a proposal for terminology and classification, *Boreas*, 3, 110-126.

Mangerud, J. and Gulliksen, S. 1975 Apparent radiocarbon ages of recent marine shells from Norway, Spitsbergen and Arctic Canada, *Quat. Res.*, 5, 263-274.

Mann, A. 1976 Ecology of Early Man in the Old World, in Ecology of the Pleistocene, *Geoscience and Man*, 14, 61-70.

Martin, P.S. 1967 Prehistoric overkill. In *Pleistocene extinctions* (eds P.S. Martin and H.D. Wright), 75-120, New Haven.

Martin, P.S. 1973 The Discovery of America, *Science*, 179, 969-974.

Mathews, R.K. 1972 Dynamics of the Ocean-Cryosphere System: Barbados data, *Quat. Res.*, 2, 368-373.

Menard, H.W. 1971 The Late Cenozoic History of the Pacific and Indian Ocean Basins. In *The Late Cenozoic Glacial Ages* (ed K.K. Turekian), 1-14.

Mercer, J.H. 1976 Glacial history of southernmost South America, *Quat. Res.*, 6, 125-166.

Mesollela, K.J., Mathews, R.K., Broecker, W.S. and Thurber, D.L. 1969 The astronomical theory of climatic change: Barbados data, *J. Geol.*, 77, 250-274.

Milankovitch, M. 1941 Kanon der Erdbestrahlung und seine Anwendung auf des Eiszeitproblem, *Acad. Roy. Serbe*, Spec. Ed., 133, 622 pp.

Milne, J.D.G. 1973 Mount Curl Tephra, a 230,000-year old marker bed in New Zealand, and its implication for Quaternary chronology, *N.Z. Jl. Geol. Geoph.*, 16, 519-532.

Miller, J.A. 1972 Dating Pliocene and Pleistocene strata using the
          potassium-argon and argon-40/argon-39 methods. In *Calibration of
          Hominoid Evolution* (ed W.W. Bishop and J.A. Miller), 63-73,
          Edinburgh.
Milliman, J.D. and Emery, K.O. 1968 Sea-Levels during the past 35,000 years,
          Science, 162, 1121-3.
Mitchell, G.F. 1972 The Pleistocene history of the Irish Sea, second
          approximation, *Sc. Proc. R. Dublin Soc.*, A4(13), 181-199.
Mitchell, G.F. 1973 INQUA - Its past, its present and its future, *Compte
          Rendu IX INQUA Congress*, 19-26, Christchurch, New Zealand.
Mitchell, G.F., Penny, L.F., Shotton, F.W. and West, R.G. 1973 A correl-
          ation of Quaternary deposits in the British Isles, *Geol. Soc. Lond.
          Spec. Rep.*, 4.
Moran, S.R. 1971 Glaciotectonic structures in drift. In *Till, a symposium*
          (eds R.P. Goldthwait *et al*), 127-148, Columbus, Ohio.
Morner, N.A. 1969 The Late Quaternary history of the Kattegat Sea and the
          Swedish West Coast..., *Sveriges Geol. Undersökn*, C-640, 1-487.
Morner, N.A. 1976 Eustasy and geoid changes, *Journ. of Geol.*, 84, 123-51.
Morner, N.A. and Lanser, J.P. 1974 Gothenburg Magnetic Flip, *Nature*, 251,
          208-409.
Morrison, R.B. 1964 Lake Lahontan: Geology of the southern Carson Desert
          Nevada, *U.S. Geol. Surv. Prof. Paper*, 424-D.
Morrison, R.B. 1965 Means of Time-Stratigraphic Division and Long-Distance
          Correlation of Quaternary Successions. In *Means of correlation of
          Quaternary Successions* (eds R.B. Morrison and H.E. Wright), Utah.
Morrison, R.B. 1966 Predecessors of Great Salt Lake. In *The Great Lake*
          (ed W.L. Stokes), 77-104, Salt Lake City.
Morrison, R.B. 1969 The Pleistocene-Holocene boundary, *Geol. en Mijnb.*, 48,
          363-372.
Morrison, R.B. and Frye, J.C. 1965 Correlations of the Middle and Late
          Quaternary Successions of the Lake Lahontan, Lake Bonneville, Rocky
          Mountain (Wasatch Range), Southern Great Plains and Eastern Mid-
          West Areas, *Nevada Bureau of Mines Rep.*, 9.
Nathan, H. 1953 Ein interglazier Schotter südlich Moosburg in Oberbayern mit
          Fagotia acicularis Ferussac, *Geol. Bavarica*, 19, 315-334.
Ninkovitch, D. and Shackleton, N.J. 1975 Distribution, stratigraphic position
          and age of ash layer 'L' in the Panama Basin region, *Earth Plant.
          Sci. Lett.*, 27, 20-35.
North, F.J. 1943 Centenary of the Glacial Theory, *Proc. Geol. Ass.*, 54, 1-28
O'Brien, T.P. 1939 *The prehistory of the Uganda Protectorate*, Cambridge.
Olausson, E. 1965 Evidence of climatic changes in deep-sea cores, with
          remarks on isotopic palaeotemperature analysis, *Prog. Oceanogr.*,
          3, 221-52.
Olausson, E. and Svenonius, B. 1973 The relation between glacial ages and
          terrestrial magnetism, *Boreas*, 2, 109-116.
Olsson, I.U. 1974 Some problems in connection with the evaluation of C^{14}
          dates, *Geol. Foren i Stockh. Forh.*, 96, 311-320.
Olsson, I.U. and Osadebe, F.A.N. 1974 Carbon isotope variations and fraction-
          ation corrections in 14C dating, *Boreas*, 3, 139-146.
Opdyke, N.D. 1972 Paleomagnetism of deep-sea cores, *Rev. Geoph. Space Phys.*,
          10, 213-249.
Osborne, P.J. 1972 Insect faunas of Late Devensian and Flandrian age from
          Church Stretton, Shropshire, *Phil. Trans. R. Soc.* B, 263, 327-367.
Page, N.R. 1972 On the age of the Hoxnian Interglacial, *Geol. J.*, 8, 129.
Pantin, C.F.A. 1968 *The relations between the sciences*, C.U.P., Cambridge.

Patterson, I.B. 1974 The supposed Perth Readvance in the Perth District, *Scot. J. Geol.*, 10, 53-66.

Patterson, W.S.B. 1969 *The physics of glaciers*, London.

Peacock, D.J. *et al* 1968 The geology of the Elgin District, *Mem. Geol. Surv. Scotland*.

Pearson, G.W., Pilcher, J.R. Baillie, M.G.L. and Hillam, J. 1977 Absolute radiocarbon dating using a low altitude European tree-ring calibration, *Nature London*, 270, 25-28.

Penck, A. and Bruckner, E. 1909 *Die Alpen im Eiszeitalter*, Leipzig.

Pennington, W. 1975 A chronostratigraphic comparison of Late-Weichselian and Late-Devensian subdivision, illustrated by two radiocarbon-dated profiles from western Britain, *Boreas*. 4, 157-171.

Pennington, W. and Bonny, A.P. 1970 Absolute Pollen Diagram from the British Late-Glacial, *Nature London*, 226, 871-873.

Phleger, F.B., Parker, F.L. and Peirson, J.F. 1953 North Atlantic core Foraminifera, *Repts. Swedish Deep Sea Expd.*, 7, 1-122.

Phillips, L. 1974 Vegetational History of the Ipswichian/Eemian Interglacial in Britain and continental Europe, *New Phytol.*, 73, 589-604.

Piggot, C.S. and Urry, W.D. 1942 Time relations in ocean sediments, *Bull. Geol. Soc. Am.*, 53, 1187-1210.

Poag, C.W. 1972 Correlation of Early Quaternary events in the U.S. Gulf Coast, *Quat. Res.*, 2, 447-467.

Poole, E.G. and Whiteman, A.J. 1961 The glacial drifts of the southern part of the Shropshire-Cheshire Plain, *Q. Jl. geol. Soc.*, 117, 91-130.

Porter, S.C. 1971 Fluctuations of Late Pleistocene Alpine Glaciers in Western North America. In *The Late Cenozoic Glacial Ages* (ed K.K. Turekian), 307-330.

Potts, A.S. 1971 Fossil cryonival features in central Wales, *Geogr. Annaler*, 53A, 39-51.

Price, R.J. 1973 *Glacial and fluvioglacial landforms*, Edinburgh.

Rafter, T.A. 1975 Radiometric Dating - Achievements and Prospects in the Quaternary. In *Quaternary Studies* (eds R.P. Suggate and M.M. Cresswell), Wellington, N.Z.

Ramsay, A.C. 1852 On the superficial accumulations and surface markings of North Wales, *Q. Jl. geol. Soc. Lond.*, 8, 371-376.

Reboul, H. 1833 *Geologie de la Periode Quaternaire...*, Paris.

Richmond, G.M. 1959 Report of the Pleistocene Committee, *American Commission on stratigraphic nomenclature Am. Ass. Pet. Geol. Bull.*, 43, 633-675.

Richmond, G.M. 1962 Quaternary stratigraphy of the La Sal Mountains, Utah, *U.S. Geol. Surv. Prof. Paper*, 324.

Richmond, G.M. 1972 Appraisal of the future climate of the Holocene in the Rocky Mountains, *Quat. Res.*, 2, 315-322.

Richter, J. 1961 Aufpressungsosartige Gletscherbruchrucken sudlich Cloppenburg in Oldenburg, *Z. Deit. Geol. Ges.*, 112, 369.

Rognon, P. and Williams, M.A.J. 1977 Late Quaternary Climatic Changes in Australia and North Africa: A preliminary interpretation, *Pal. Pal. Pal.*, 21, 285-327.

Rona, E. and Emiliani, C. 1969 Absolute dating of Caribbean cores P6304 and P6304-9, *Science*, 163, 66-68.

Rose, J. and Allen, P. 1977 Middle Pleistocene stratigraphy in south east Suffolk, *Jl. geol. Soc. Lond.*, 133, 83-102.

Ruddiman, W.F., Sancetta, C.D. and McIntyre, A., 1977 Glacial/Interglacial response rate of subpolar North Atlantic waters to climatic change: the record in oceanic sediments, *Phil. Trans. R. Soc.* B, 280, 119-142.

Ruhe, R.V. 1965 Quaternary palaeopedology. In *The Quaternary of the United States* (ed H.E. Wright and D.G. Frey), 755-764, Princeton.

Runge, E.C.A., Goh, K.M. and Rafter, T.A. 1973 Radiocarbon chronology and
    problems in its interpretation for Quaternary loess deposits -
    South Canterbury, New Zealand, *Proc. Soil Sci. Soc. Am.*, 37, 472-476.
Sancetta, C., Imbrie, J., Kipp, N.G., McIntyre, A. and Ruddiman, W.F. 1972
    Climatic record in North Atlantic Deep-Sea core V23-82: Comparison
    of the Last and Present Interglacials Based on Quantitative Time
    Series, *Quat. Res.*, 2, 363-367.
Savage, D.E. and Curtis, G.H. 1967 The Villafranchian Age and its radiometric
    dating, *Bull Am. Assoc. Pet. Geol.*, 51, 479.
Savin, S. and Douglas, R.G. 1973 Oxygen isotope and magnesium geochemistry of
    recent planktonic Foraminifera from the South Pacific, *Bull. geol.
    Soc. Am.*, 84, 2327-42.
Schott, W. 1935 Die Foraminiferen. In *Wissenschaftliche Ergebnisse der
    Deutschen Atlantischen Exp 'Meteor'*, 1925-1927, 3, Berlin.
Schafer, I. 1953 Die donaueiszeitlichen Ablagerungen en Lech und Wertach,
    Geol. Bavarica, 19, 13-54.
Schofield, J.C. 1960 Sea-level fluctuations during the past four thousand
    years, *Nature*, 185, 836.
Schumm, S.A. 1965 Palaeohydrology. In *The Quaternary of the United States*
    (eds H.E. Wright and D.G. Frey), Princeton.
Schwarzbach, M. 1963 *Climates of the Past*, London.
Selle, W. 1962 Geologische und vegetationskundliche Untersuchungen an eingen
    wichtigen Vorkommem des letzen Interglazials in Nordwestdeutschland,
    Geol. Jahrb., 79, 295.
Shackleton, N.J. 1967 Oxygen isotope analyses and Pleistocene temperatures
    re-assessed, *Nature, Lond.*, 215, 15-17.
Shackleton, N.J. 1968 Depth of pelagic Foraminifera and isotope changes in
    Pleistocene oceans, *Nature Lond.* 218, 79-80.
Shackleton, N.J. 1969 The last interglacial in the marine and terrestrial
    record, *Proc. R. Soc. Lond.* B 174, 135-154.
Shackleton, N.J. 1974 Attainment of isotopic equilibrium between ocean water
    and the benthonic foraminifera genus *Uvigerina*: isotopic changes
    in the ocean during the last glacial, *Colloques Int. Cent. natn.
    Rech. Scient.*, 219, 203-10.
Shackleton, N.J. 1975 The Stratigraphic Record of Deep-Sea Cores and its
    implications for the assessment of Glacials, Interglacials and
    Interstadials in the Mid-Pleistocene. In *After the Australopithecines*,
    (ed K.W. Butzer and G.Ll. Isaac), 1-24, The Hague.
Shackleton, N.J. 1977a Oxygen isotope stratigraphy of the Middle Pleistocene,
    In *British Quaternary Studies* (ed F.W. Shotton), 1-16, Oxford.
Shackleton, N.J. 1977b The oxygen isotope record of the Late Pleistocene,
    Phil. Trans. R. Soc. B 280, 169-182.
Shackleton, N.J. and Opdyke, N.D. 1973 Oxygen isotope and palaeomagnetic
    stratigraphy of Equatorial Pacific core V28-238: oxygen isotope
    temperatures and ice volumes on a 10^5 ycar and 10^6 year scale,
    Quat. Res., 3, 39-55.
Shackleton, N.J. and Opdyke, N.D. 1976 Oxygen isotope and palaeomagnetic
    stratigraphy of Equatorial Pacific core V28-239, Late Pliocene to
    Latest Pleistocene. In *Investigation of late Quaternary palaeo-
    oceanography and palaeoceanography* (ed R.M. Cline and J.D. Hays),
    Mem. geol. Soc. Am., 145, 449-64.
Shephard, F.P. 1973 *Submarine Geology*, New York.
Shotton, F.W. 1953 The Pleistocene deposits of the area between Coventry,
    Rugby and Leamington and their bearing upon the topographical
    development of the Midlands, *Phil. Trans. R. Soc.* B, 237, 209-260.

Shotton, F.W. 1967a The problems and contributions of methods of absolute dating within the Pleistocene period, *Jl. geol. Soc. Lond.*, 122, 357-83

Shotton, F.W. 1967b Age of the Irish Sea Glaciation of the Midlands, *Nature London*, 215, 136.

Shotton, F.W. 1968 The Pleistocene succession around Brandon, Warwickshire, *Phil. Trans. R. Soc.*, 254, 387-400.

Shotton, F.W. 1972 An example of hard water error in radiocarbon dating of vegetable matter, *Nature London*, 240, 460-1.

Shotton, F.W. 1976 Amplification of the Wolstonian Stage of the British Pleistocene, *Geol. Mag.*, 113, 241-250.

Shotton, F.W. 1977a British dating work with radioactive isotopes. In *British Quaternary Studies* (ed F.W. Shotton), 17-29, Oxford.

Shotton, F.W. 1977b Chronology, Climate and Marine Record, The Devensian Stage ..., *Phil. Trans. R. Soc.* B 280, 107-118.

Sissons, J.B. 1961 The central and eastern parts of the Lammermuir-Stranraer moraine, *Geol. Mag.*, 98, 380-392.

Sissons, J.B. 1966 Relative sea-level changes between 10,300 and 8300 B.P. in part of the Carse of Stirling, *Trans. Inst. Br. Geogr.*, 48, 39, 19-29.

Sissons, J.B. 1974 Lateglacial marine erosion in Scotland, *Boreas*, 3, 41-48.

Sissons, J.B. 1976 *Scotland: The geomorphology of the British Isles*, London.

Stearns, C.E. 1976 Estimates of the position of sea level between 140,000 and 75,000 years ago, *Quat. Res.*, 6, 445-450.

Stearns, C.E. and Thurber, D.L. 1965 Th^{230}/U^{234} dates of late Pleistocene marine fossils from the Mediterranean and Moroccan littorals, *Quaternaria*, 7, 29-42.

Steinen, R.P., Harrison, R.S. and Mathews, R.K. 1973 Eustatic low stand of sea level between 125,000 and 105,000 years BP: evidence from the subsurface of Barbados, West Indies, *Bull. Geol. Soc. Am.*, 84, 63-70.

Straw, A. 1969 Pleistocene events in Lincolnshire: a survey and revised nomenclature, *Trans. Lincs. Nat. Union*, 17, 85-98.

Stuart, A.J. 1974 Pleistocene history of the British vertebrate fauna, *Biol. Rev.*, 49, 225-66.

Stuart, A.J. 1976 The history of the mammal fauna during the Ipswichian/Last Interglacial in England, *Phil. Trans. R. Soc.* B 276, 221-50.

Stuiver, M. 1970 Long term C14 variations. In *Radiocarbon variations and Absolute Chronology* (ed. I. Olsson), 197-214, Stockholm.

Suess, H.E. 1965 Secular variations of the cosmic ray produced carbon-14 in the atmosphere, *J. Geophys. Res.*, 70, 5937-5952.

Sugden, D.E. 1970 Landforms of deglaciation in the Cairngorm Mountains, *Trans. Inst. Br. Geogr.*, 51, 201-219.

Suggate, R.P. 1965 The definition of "Interglacial", *J. Geol.*, 73, 619-626.

Suggate, R.P. 1974 When did the Last Interglacial end? *Quat. Res.*, 4, 246-252.

Sutcliffe, A.J. 1960 Joint Mitnor Cave, Buckfastleigh, *Trans. Proc. Torquay nat. hist. Soc.*, 13, 1-26.

Szabo, B.J. and Rosholt, J.N. 1969 Uranium-series dating of Pleistocene molluscan shells from Southern California - an open system model, *J. Geophys. Res.*, 74, 3253-3260.

Szabo, B.J. and Collins, D. 1975 Ages of fossil bones from British interglacial sites, *Nature Lond.*, 254, 680-82.

Tauber, H. 1970 The Scandinavian varve chronology and C14 dating. In *Radiocarbon variations and Absolute Chronology* (ed I. Olsson), 173-196, Stockholm.

Terasmae, J. 1960 Contributions to Canadian palynology, *Canadian Geol. Surv.*, 56, 41 pp.

Thomas, G.S.P. 1976 The Quaternary stratigraphy of the Ice of Man, *Proc. Geol. Assoc.*, 87.

Thom, B.G. 1973 The dilemma of high interstadial sea levels during the last glaciation, *Prog. in Geogr.*, 5, 170-245.

Thorarinsson, S. 1954 The tephra-fall from Hekla on March 29th 1947. In *The Eruption of Hekla 1947-1948* (eds. T. Erinarsson, T., Kjartansson, G. and Thorarinsson, S.), 1-68 (Part 2). Reykjavik.

Thurber, D.L. 1972 Problems of dating non-woody material from continental environments. In *Calibration of Hominoid Evolution* (eds. W.W. Bishop and J.A. Miller), 1-18, Edinburgh.

Thurber, D.L., Broecker, W.S., Blanchard, R.L. and Potratz, H.A. 1965 Uranium-series ages of Pacific atoll coral, *Science*, 149, 55-58.

Tooley, M.J. 1974 Sea level changes during the last 9000 years in north west England, *Geog. J.*, 140, 18-42.

Turekian, K.K. and Bada, J.L. 1972 The dating of fossil bones. In *Calibration of Hominoid Evolution* (eds W.W. Bishop and J.A. Miller), 177-186, Edinburgh.

Turner, C. 1970 The Middle Pleistocene deposits at Marks Tey, Essex, *Phil. Trans. R. Soc.* B 257, 373-440.

Turner, C. 1975 The correlation and duration of Middle Pleistocene inter-glacial periods in Northwest Europe. In *After the Australopithecines* (ed. K.W. Butzer and G.Ll. Issac), 259-308, Mouton, The Hague.

Turner, C. 1977 In Guidebook to Excursion A1/C1 East Anglia, *INQUA X CONGRESS* (ed D.Q. Bowen).

Turner, C. and West, R.G. 1968 The subdivision and zonation of interglacial periods. *Eiszeit. u. Gegenw.*, 19, 93-101.

Urey, H.C. 1947 The thermodynamic properties of isotopic substances, *J. Chem. Soc.*, 562-581.

Valentine, K.W.G. and Dalrymple, J.B. 1976 Quaternary Buried Palaeosols: A Critical Review, *Quat. Res.*, 6, 209-222.

Velichko, A. 1969 Les traits essentiels de la stratigraphie des loess de la plain d'Europe orientale. In *La Stratigraphie des Loess d'Europe, Bull Assoc. Fr. Etude Quat. Suppl.*, 160-164.

Vita-Finzi, C. 1973 *Recent earth history* (Macmillan), London.

Vries, de H. 1958 Variation in concentration of radiocarbon with time and location on earth, *Koninkl. Nederlands Akad. Wetensch. Proc.* ser. B, 61, 1-9.

Vucetich, C.G. and Pullar, W.A. 1964 Stratigraphy and chronology of the late Quaternary volcanic ash in Taupo, Rotorua and Gisborne districts, *N.Z. geol. Surv. Bull.*, 73, 88 pp.

Walcott, R.L. 1970 Isotatic response to loading of the crust in Canada, *Canadian Journ. of Earth Sciences*, 7, 716.

Walcott, R.I. 1972 Past sea levels, Eustasy and Deformation of the Earth, *Quat. Res.*, 2, 1-14.

Walcott, R.I. 1973 Structure of the Earth from 1000l Glacio-Isotatic rebound. *Annual Review of Earth and Planetary Sciences*, vol. 1, 15-37.

Washburn, A.L. 1971 Quaternary Research and Education, *Quat. Res.*, 1, 283-284.

Washburn, A.L. 1973 *Periglacial Processes and Environments*, London.

Watkins, N.D. 1971 Geomagnetic Polarity Events and the Problem of "The Reinforcement Syndrome", *Comments on Earth Sciences - Geophysics*, 2, 36-43.

Watkins, N.D. 1972 Review of the development of the Geomagnetic Polarity time scale and discussion of prospects for its finer definition, *Bull. Geol. Soc. Am.* 83, 551-574.

Watkins, N.D. 1976 Polarity Commission sets up some guidelines, *Geotimes*, 21(4), 18-20.

Watson, E. 1972 Pingos of Cardiganshire and the latest ice limit, *Nature London,* 238, 343-344.

Watson, E. 1977 Mid and North Wales, *X INQUA CONGRESS*, Excursion guidebook (ed. D.Q. Bowen).

Watts, W.A. 1974 The vegetation record of a mid-Wisconsin Interstadial in northwest Georgia,*Quat. Res.,* 3, 257-268.

Wayland, E.J. 1934 Rifts, rivers, rains, and early man in Uganda, *Roy. Anthropol. Inst. J.,* 333-352.

Weeks, A.G. 1969 The stability of natural slopes in south-east England as affected by periglacial activity, *Q.Jl. Engng. Geol.,* 2, 49-61.

Weertman, J. 1961 Stability of ice age ice sheets, *J. Geophys. Res.,* 66, 3788-3792.

Wendland, W.M. and Donley, D.L. 1971 Radiocarbon Calendar age relationship, *Earth Planet. Sci. Lett.,* 11, 135-139.

West, R.G. 1956 The Quaternary deposits at Hoxne, Suffolk, *Phil. Trans. R. Soc.* B 239, 265-356,(A).

West, R.G. 1957 Interglacial deposits at Bobbitshole, Ipswich, *Phil. Trans. R. Soc.* B 241, 1-31.

West, R.G. 1971 *Studying the past by pollen analysis,* Oxford.

West, R.G. 1973 In discussion with Bristow and Cox *op cit.*

West, R.G. 1975 *Pleistocene geology and biology*, London.

West, R.G. and Donner, J.J. 1956 The glaciations of East Anglia and the East Midlands: a differentiation based on stone orientation measurements of the till, *Q. Jl. geol. Soc.,* 112, 69-91.

West, R.G. and Wilson, D.G. 1966 Cromer Forest Series, *Nature London,* 209, 497-498.

Wiegank, K.F. 1972 Okologische Analyse quartarer Foraminiferen, *Geologie,* 21, 1-111, Berlin.

Wijmstra, T.A. 1969 Palynology of the first 30 metres of a 120 m deep section in northern Greece, *Acta Botan. Neerl.,* 18, 511.

Wijmstra, T.A. and Hammen, van der T., 1974 The Last Interglacial-Glacial cycle: state of affairs of correlation between data obtained from the land and from the ocean, *Geol. en Mijnb.,* 53, 386-392.

Williams, J. 1975 The influence of snow cover on the atmospheric circulation and its role in climatic change: an analysis based on results from the NCAR global circulation model, *J. App. Met.,* 14, 137-152.

Willman, H.B. and Frye, J.C. 1970 Pleistocene Stratigraphy of Illinois, *Illinois Geol. Surv. Bull.,* 94.

Woillard, G. 1975 Recherches palynologiques sur le Pleistocene dans l'est de la Belgique et dans les Vosges Lorraines, *Acta Geogr. Lovaniensia,* 14, 118 pp.

Woldstedt, P. 1951 *Das Eizeitalter, Grundlinien einer Geologie des Quarters,* Vol. 1 (Vol. 2 1958), Stuttgart.

Woldstedt, P. 1967 The Quaternary of Germany. In *The Quaternary* (ed K. Rankama), 239-300, New York and London.

Worsley, P. 1967 Problems in naming the Pleistocene deposits of the North East Cheshire Plain, *Mercian Geol.* 2, 51-55.

Worsley, P. 1976 Correlation of the Last Glaciation maximum and the extra-glacial fluvial terraces in Britain - a case study. In *Quaternary Glaciations in the Northern Hemisphere* (eds D.J. Easterbook and V. Sibrava), 274-284, Prague.

Worsley, P. 1977 In Wales and the Cheshire-Shropshire Lowland Guidebook for excursion A8/C8 *X INQUA CONGRESS* (ed D.Q. Bowen).

Wright, H.E. 1973 Quaternary Science and Public Service, *Quaternary Research,* 3, 515-519.

Wright, H.R. 1976a Pleistocene ecology - some current problems, in Ecology
        of the Pleistocene, *Geoscience and Man*, 13, 1-12.
Wright, H.R. 1976b The dynamic nature of Holocene vegetation. A problem in
        paleoclimatology, Biogeography and Stratigraphy nomenclature,
        Quat. Res., 6, 581-596.
Wright, W.B. 1937 *The Quaternary ice age*, London.
Wymer, J.J., Jacobi, R.M. and Rose, J. 1975 Late Devensian and Early
        Pleistocene barbed points from Sproughton, Suffolk, *Proc. prehist.
        Soc.*, 41, 235-41.
Zagwijn, W.H. 1974 The Pliocene-Pleistocene boundary in western and southern
        Europe, *Boreas*, 3, 75-97.
Zagwijn, W.H. 1975 Variations in climate as shown by pollen analysis
        especially in the Lower Pleistocene of Europe. In *Ice ages: ancient
        and modern* (eds A.E. Wright and F. Moseley), (Seel House, Liverpool),
        137-52.
Zagwijn, W.H., Montfrans, van H.M., and Zandstra, J.G. 1971 Subdivision of
        the "Cromerian" in the Netherlands: pollen analysis, palaeomagnetism
        and sedimentary petrology, *Geol. en Mijnb.*, 50, 41-58.
Zeuner, F.E. 1946 *Dating the Past*, London.
Zeuner, F.E. 1959 *The Pleistocene Period*: its climate, chronology and faunal
        successions, 2nd edn., London.

ADDENDA

Berggren, W.A. and Couvering, J.A. van 1974 The Late Neogene: Biostrati-
        graphy, geochronology and palaeoclimatology of the last 15 million
        years in marine and continental sequences, *Pal. Pal. Pal.*, 16,
        1-216.
Richmond, G.M. 1976 Pleistocene stratigraphy and chronology in the mountains
        of Western Wyoming. In *Quaternary Stratigraphy of North America*
        (ed W.C. Mahaney), 353-379, Stroudsburg, Penn.
Rosholt, J.N., Emiliani, C., Geiss, J., Koczy, F.F., and Wangersky, P.J.
        1961 Absolute dating of deep-sea cores by the Pa^{231}/Th^{230} method,
        J. Geol., 69, 162-185.
Ruhe, R.V. 1976 Stratigraphy of mid-continent loess in U.S.A. In
        Quaternary Stratigraphy of North America (ed W.C. Mahaney), 197-
        212, Stroudsburg, Penn.
Sutcliffe, A.J. 1975 A hazard in the interpretation of glacial-interglacial
        sequences, *Quat. News*, 17, 1-3.
Sutcliffe, A.J. 1976 reply to R.G. West, *Quat. News*, 18, 1-7.
Sutcliffe, A.J. and Kowalski, K. 1976 Pleistocene rodents of the British
        Isles, *Bull. Br. Museum*, 27(2), 147 pp.

INDEX